POLICE HELICOPTER

Delivering air support for law enforcement

COVER IMAGE: Two of the Metropolitan Police's Eurocopter EC145 helicopters pictured in a rare formation flight over London. The Shard building seen in the background rises to a height of 1,016ft above the capital. *(Author's collection)*

First published in March 2020

A catalogue record for this book is available from the British Library.

ISBN 978 1 78521 570 4

Library of Congress control no. 2019945934

Published by Haynes Publishing,
Sparkford, Yeovil, Somerset BA22 7JJ, UK.
Tel: 01963 440635
Int. tel: +44 1963 440635
Website: www.haynes.com

Haynes North America Inc.,
859 Lawrence Drive, Newbury Park,
California 91320, USA.

Printed in Malaysia.

Senior Commissioning Editor: Jonathan Falconer
Copy editor: Michelle Tilling
Proof reader: Penny Housden
Indexer: Peter Nicholson
Page design: James Robertson

Acknowledgements

A great number of people have helped and supported me in gathering the photos and information needed to tell the story of police aviation.

I would like to start by thanking all the police officers, pilots, engineers and staff that I am proud to have worked with during my years with the Metropolitan Police ASU. As the Accountable Manager and Unit Executive Officer I relied heavily upon the support of my management team and had so much fun working with them to gain the knowledge and create the memories that have gone into this book. My thanks go to Sergeant John Gleeson (retd), Sergeant Andy Hutchinson (retd) and Sergeant Terry White (retd), Captain Nick West, Captain Kevin Sutton, Captain Paul Watts and Chief Engineer Roger Stevenson, who formed a special team that delivered excellence in air support for London.

Special thanks need to go to the NPAS staff still serving at NPAS London, who always make me so very welcome when I visit Lippitts Hill and who answer each and every question or challenge I put to them with excellent results. Special thanks go to Sergeant Dan Arnold, Captain John Roberts, Constable Bev MacWilliam, Constable Hugh Dalton and Constable Colin Barker, who helped me greatly with the material for this book.

Thanks to Inspector Phil Whitelaw (retd) and Sergeant Tony Mepham (retd) from whom I learned so much about the formation and function of air support. To Bryn Elliott who is a fount of all knowledge when it comes to the history of police aviation and who supported me with a vast amount of detailed information and photographs. I am sure there are many more people to thank, so apologies to anyone I have forgotten to mention.

Thank you to my wife and family who have always supported me as I have indulged in my passion for police helicopters. Your tireless love and support are really appreciated, and I hope that when you read this book you will understand why I am so passionate about this subject.

Finally, to all the men and women who deliver air support on behalf of NPAS today, you should be rightfully proud of the service you provide to help fight crime and maintain the safety of the public you serve and police officers you support.

This book is dedicated to the memory of the brave men and women who have lost their lives while delivering air support. Thank you for your service.

POLICE HELICOPTER

Delivering air support for law enforcement

Operations Manual

Insights into police aviation in Great Britain, the Metropolitan Police Air Support Unit, police helicopter aircrews, equipment and tactics

Richard Brandon

Contents

OPPOSITE The Metropolitan Police's EC145 helicopter, G-MPSC, is pictured in the hover over Parliament Square during a public order event in Westminster in June 2013, filmed from an accompanying police helicopter that is downlinking the action live into the control room. *(All photos from author's collection unless credited otherwise)*

Introduction

The best way to introduce this book is to start by explaining what qualifies me to write it and then how I came to do so. My involvement with Police aviation started in 2000 when as a Metropolitan Police sergeant based at Wembley Police Station I answered an internal advert inviting applicants for the role of air support unit (ASU) sergeant. Some 12 months later, after a rigorous selection process, PS John Gleeson and I were selected for the role. The only problem was that there was only one vacancy, and John was posted into it. I had to wait a further 12 months for my opportunity: eventually in mid-2002 a vacancy came up, and my air support career started.

My passion for aviation did not really emerge until after I arrived at the ASU. I had not flown in a helicopter until my flight test assessment and had no aviation background or experience. My police career had begun some 15 years earlier when I joined the Met as a PC, moving to London from Durham in the north-east of England. While completing my A levels I had seen a newspaper advert for the role of police constable in London's Metropolitan Police. Policing seemed an exciting and interesting career, and the sheer size and scale of the Met meant that an application from me as an 18-year-old was considered. In February 1988, shortly after my 19th birthday, I was in training at the Metropolitan Police Training School in Hendon – and so started my 30-year policing career. Like almost all tactical flight officers (TFOs) in air support today, I started as an operational PC policing the streets, and it was while working on West Hendon Division in 1989 that I first became aware of air support. I witnessed 'India 99', the Met's Bell 222 helicopter, assisting with a suspect search in Colindale, and remember asking one of the experienced PCs how you came to be posted to the police helicopter. He told me it was jobs for the boys and there was

no point in looking at the role; so I gave it no further thought.

Winding the clock forwards 11 years, I had completed a posting as an instructor at the recruit training school, had been promoted to sergeant and was in the event planning office at Wembley Police Station when I read the advert for the ASU sergeant role. I remember reading through the essential criteria and thinking 'I could do that', but then dismissing the idea as I felt there was no point in applying. PS Andy Hooper, whom I shared an office with, turned to me and said, 'Why not give it a go? What have you got to lose if you don't get it?' He was right, and in spring 2002 I was posted to the ASU to work alongside Inspector Phil Whitelaw and Sergeants John Gleeson and Dick Pooley. Dick was soon to retire, and we were joined by Sergeant Andy 'Hutch' Hutchinson. The team that would eventually run air support was in place.

Owing to my training and assessment background and qualifications I was asked to be training manager, a role that, after a year of learning the TFO ropes, I was delighted to take on. My time at Lippitts Hill would see me designing, writing and delivering TFO selection and training. I also oversaw the mid-life upgrade of the three AS355N Twin Squirrel helicopters, which added the Wescam MX15 and Skyforce mapping to Met helicopters for the first time. The day-to-day running of the ASU was challenging but enjoyable, and I quickly realised I was a round peg in a round hole. My promotion ambitions vanished, and I decided that I would stay in air support as long as the Met would let me. I picked up the nickname 'Brains' from my then Chief Superintendent Simon Humphrey, who in a speech at the unit's 25th anniversary dinner described the members of the air support team as being like characters from *Thunderbirds*. I know nicknames are not really acceptable in policing today, but I was proud of mine and it stuck for many years.

In 2005 I was given a massive project to

oversee. This was the introduction into service of a fleet of three new helicopters. The EC145 was a larger and more capable helicopter than the AS355N, and it was to be fitted with role equipment in such a way that it could be re-roled to undertake a variety of passenger transport missions in addition to general policing. I spent the next 18 months working to get the design and role fit of the new helicopters right, writing and completing acceptance testing and then delivering them into operational service, something that happened in July 2007. The focus from 2007 onwards was the Olympic Games in London in 2012: we had to be ready with a full range of operational tactics that had not previously been undertaken by police air support. I continued to fly as a TFO when I could and trained to be a fast rope dispatcher, a role that was unlike anything I had ever done before and great fun.

The Olympic Games passed without incident, and shortly afterwards Inspector Phil Whitelaw announced his retirement after no less than 40 years of police service. Having qualified as an inspector and been selected for promotion, everything aligned to see me retained in a role I loved – and I was promoted to inspector as head of the ASU and Accountable Manager. My passion for air support has never faded, and although my time in the area came to an end when the Met joined the National Police Air Service (NPAS) in 2015 I was lucky enough to end my policing career on the Met's Taskforce. I spent my last three years of service working with the talented professionals who make up the Mounted Branch, Dog Section, Marine Policing Unit and Territorial Support Group. Taskforce is an amazing command with dedicated and professional officers who are all specialists in their own right. Air support has always, however, defined my career, and at my retirement function it was joked that I might have said 'I used to be on air support, you know' to Taskforce colleagues a few times. My service as a police officer in London is my proudest achievement, but I will always love police aviation and am proud to have been part of it for almost half of my police career.

So how come I am writing a Haynes manual? I remember encountering these iconic manuals when I was a teenager trying to fix my

Vespa scooter or my dad's car. I had no idea that Haynes had begun producing manuals on engineering landmarks such as the RAF Typhoon, RAF Tornado or indeed the London Underground until I was bought one as a birthday present. I started to ask myself if a manual about police helicopters might work. Throughout my time in air support I had always been amazed by the response we received when we opened the door to the public and showed them inside our world. We had filmed two series of Sky Cops for the BBC and pioneered the use of social media (specifically Twitter and Instagram) to promote and explain what we were doing. The interest the public had shown in air support was fantastic: whenever we landed in a park or at an event we were always inundated with questions about the helicopter, its equipment and its crew. I have never seen a book like this and I have lots of material that I hope you will enjoy – so here it is.

I hope you will consider me qualified to write this manual and that you enjoy this insight into a specialised area of policing, one that I believe brings so much value to police officers and the public we serve. This book lifts the lid on police aviation, and I hope it will inspire the next generation of police officers, police pilots and TFOs.

LEFT The author pictured during open-door photography through the rear sliding door of the EC145. Whenever the rear sliding door is open, a fall arrest harness attached to the helicopter is worn in addition to the Martin-Baker seat harness. The seat can be rotated for comfort to face the open door.

Chapter One

Police aviation in the UK

Police officers have always used every means available to stay ahead of criminals. In the early 20th century they first took to the skies to exploit the advantages that an aerial vantage point would give them, and police air support was born. These early trials laid the foundations of the essential operational tactic taken for granted in policing today.

OPPOSITE **The South Wales EC135 helicopter over the Millennium Stadium in Cardiff.** *(Gary Smart)*

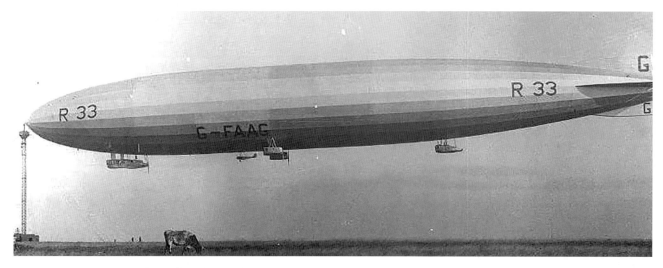

In the beginning

Police officers have always been resourceful when it comes to the fight against crime. The earliest example of this resourcefulness that involved aviation took place in the United States in 1914, when police officers commandeered a Curtiss Model F flying boat to give chase to a jewellery thief. Little is known about this operation, which was apparently successful, but it demonstrates that for well over a century police officers have been seeking to gain an airborne advantage over the criminals they pursue.

The first experimental police flight in the UK is reputed to have taken place at the annual Epsom and Ascot summer race meetings, in south-west London in May and June 1920. The aircraft involved didn't have radio technology fitted and had to land near the racecourse to pass on observations. In the post-war years, the chief constables of police forces across the UK were typically ex-military men who had served as commanding officers in the Army or Royal Air Force. It is no surprise therefore that they considered the use of aviation in policing to be a viable proposition, and that they supported trials and experiments. By 1921 the Metropolitan Police had decided it would experiment, this time with an R33 airship over the Epsom Derby to manage the traffic attending the event.

In 1922 the Metropolitan Police recorded its first use of a fixed-wing aircraft, a Vickers Type 61 Vulcan, to undertake traffic control duties at the Derby. The Vulcan biplane was not the most appropriate type of aircraft, as it had a fully enclosed cabin and small windows, with the view from these being obstructed by an array of struts and wires linking the biplane's wings. The experiment was reported as a success, although the absence of any further developments by the Met perhaps suggests otherwise.

During the 1930s developments began to quicken. Leicestershire Police, guided by their ex-RAF chief constable, used a borrowed de Havilland Moth for crime detection rather than just traffic control for the first time. The Met began to use an autogyro flown from RAF Hendon and two flights were set aside to investigate whether it was feasible for the

aircraft to follow a car from Hendon to central London. The first attempt was so successful that it was repeated to ensure it was not a fluke! Areas in which the autogyro excelled were the inspection of enclosed premises and the searching of open spaces, but despite the success of these tests it was only many years later that a new generation of police fliers took to the skies – when the experiments ended with the outbreak of war in 1939.

Post-Second World War

After the war, in 1946, efforts began again to bring about growth in police aviation. Aircraft manufacturers made a variety of offers to the police service, and Lancashire Constabulary funded an aircraft to provide air support over the Aintree racecourse for the Grand National. The country was rebuilding

after the war, food, money and resources were scarce and many police forces were struggling to recruit the resources they needed for policing, let alone for aviation. In April 1949, however, it appears that a twin-engine Miles M65 Gemini carried a police air observer with a radio set over the racecourse on Grand National Day. Poor aircraft reliability and unsatisfactory cabin design were encountered during trials, with the main issue being poor communications in a noisy cockpit. At the time there was no standard for police radio communications, and officers on flying duties brought with them a bulky Marconi 'walkie-talkie' radio set. The idea of a fully role-equipped police helicopter, like those found in service today, was still a long way off.

In 1955 the Metropolitan Police used a Bristol helicopter to observe traffic approaching Wembley Stadium for the FA Cup Final fixture

ABOVE Aerial photo of Wembley Stadium taken by a police air observer in 1956. The image shows the car parks and thousands of football fans walking around the stadium. *(Brynn Elliott, Air Support Archive)*

between Manchester City and Birmingham. A chief inspector acted as air observer over five flights, which is in stark contrast to the police constable and police sergeant tactical flight officers (TFOs) who fly today. It was noted how easily the helicopter, from 1,000ft above ground, was able to keep the slow moving royal party in view and monitor the movement of traffic and crowds.

Trials continued in a number of police forces, and by the mid-1950s Bristol had teamed up with Pye of Cambridge Ltd, manufacturers of television equipment, to investigate the possibilities of broadcasting television pictures from a Bristol 171 helicopter to a ground station. This capability was to become a key requirement of future police helicopter operations, with imagery from the onboard camera systems being distributed into police

control rooms. The trials used a small hand-held camera linked to a portable monitor in the helicopter, and pictures were transmitted successfully to the ground for viewing on a 14in domestic TV. The concept of heli-tele was born. However, early heli-tele turrets were huge and very heavy, and their black and white imagery was far from clear and reliable.

By 1957 the Met was trialling the use of air support again, and this time chose to use an Auster 5 aircraft operated by Croydon-based Vendair. The aim was to observe bank holiday traffic and crowds. Unfortunately, the crew encountered a number of problems, primarily caused by the impact of cockpit noise upon essential communications: the radio telephone (RT) sets in use couldn't cope with the cockpit noise levels. The Home Office approved the continuation of air support in 1958, and from

this point an aircraft-mounted radio set was used in preference to the hand-held radios. The decision was made to fly a single police air observer who needed to be a trained RT operator. The selection criteria for air observers was very specific, requiring low-level flying experience and the ability to identify targets from the air. Eventually two sergeants and an inspector, all of whom were proven radio operators, were chosen for the role.

The 1960s

In 1960 a meeting of the Central Conference of Chief Constables discussed air support. It was suggested that three well-placed machines would be able to cover the whole of England and Wales. The proposal was referred to committee, but a year later a report to the next conference admitted that no significant progress had been made.

In the early 1960s British Executive Aircraft Services (BEAS), based at Oxford Airport, Kidlington, handled the UK franchise for the American Brantly B2 series of light helicopters, which was about the smallest and simplest two-seat helicopter that was still usable. An offer was accepted by the chief constable of Durham Constabulary to try out this helicopter in the north-east of England. The trials were wide ranging, but frequently traffic related: observers were instructed to log and report upon what was seen on motorways. They also included escorting cash in transit and working with police dogs and handlers to pursue and

detain suspects. The BEAS Brantly, registration G-ARYX, was equipped with a light closed box with a forward-facing mesh door that was hung on its port side slightly to the rear of the passenger seat. A single Alsatian police dog was carried in this kennel, which was easily accessible to the police dog handler in the cabin. When the helicopter came to a low hover, at the appropriate moment the handler could release the door and allow the dog to leap out. To assist with this, the door was removed from the passenger side of the cabin. One can only imagine what the Health and Safety Executive let alone the Civil Aviation Authority (CAA) would have to say about such plans today.

Work continued to address the communication issues that crews encountered when undertaking police air support work.

ABOVE Photo pilot and observer briefing next to Vendair-owned Auster 5 G-AKXP in 1958. *(Brynn Elliott, Air Support Archive)*

LEFT A BEAS Brantly G-ARYX was used for trials by the police in the 1960s. These included the carriage of police dogs in a container attached to the outside of the helicopter. *(Mike Hines, Air Britain Photographic Images Collection)*

Through careful mounting of the antenna and improvements to the headset and microphone equipment, many of the earlier problems were solved. An attempt was made to provide the crew with a number of transmissions simultaneously. All police transmissions were received in the observer's left ear and intercom input from the pilot, air traffic control (ATC) and his own throat mike was carried to the right. This was far cruder than the sophisticated modern system that overlays all radio and intercom sources to both ears, but it worked reasonably well. By the late 1960s trials were using AAC Bell 47 Sioux helicopters drawn from a military reserve pool. These had a police radio and some additional 'POLICE' door signs as the only extra equipment. They were operated across the UK and took on various policing roles, most notably the search for prisoners who escaped from Dartmoor Prison around Christmas in 1966. Aerial observation from helicopters and the transportation of five officers in a larger AAC Westland Scout AH1 proved the benefits of rapid air transportation of specialist resources across the country.

BELOW The police used Army Air Corps Bell 47 Sioux helicopters for air support trials. The extent of the role equipment was a police sign and carry on radio. *(Brynn Elliott, Air Support Archive)*

During the London phase of these trials, in July 1967, Lippitts Hill, the current location of NPAS London, was first used. This former military site was ideally located to the north-east of the capital and was within easy flying time of central London.

Into the 1970s

In 1971 the Metropolitan Police selected a number of air observers from among the Met's best 'thief takers' for the Air Support Unit (ASU). One such observer was PC Phil Snoad, who recounts his adventures as one of the first Met ASU observers in his book *Eye in the Sky*. The team of observers was small, and its members flew a number of missions in a range of helicopters that included the Hughes 269-300. It was at this point that the ASU first adopted the famous call sign 'India 99' for its communications with the Met's control room, which had the call sign 'M2-MP' – or 'MP' for short.

In the meantime air support use by police around the world continued to develop, with new and more powerful helicopters, such as

the twin-engine Bölkow Bo 105, the Bell 206B JetRanger and Aérospatiale AS365C Dauphin, being used. At the time these aircraft were considered to be expensive, and there were many arguments between police chiefs and politicians over funding – and whether there was a real need to spend so much money on air support. However, the twin-engine aircraft would eventually become the staple of UK policing, and remains in use around the world today. A further change at this time involved the helicopter supplier contract, which saw the introduction of the Sud SA341 Gazelle into police operations alongside the Hughes 269.

The technology that was to become the beating heart of any modern police helicopter was also developing, with the Army Air Corps (AAC) using the new heli-tele in the policing of Ireland during the Troubles. The new heli-tele performance was breathtaking and in testing early examples met or exceeded the aims of the AAC requirement. The secret of the success of heli-tele was primarily the combination of camera and lens with gyro-stabilisation. The Phillips LDK 14D featured

a 25:1 zoom facility and a 1.5 times range extender, which could zoom down from 20° to only 1°. What this meant was that the AAC Scout helicopters were able to operate away from the target area virtually unnoticed, yet take in every detail of the action.

The late 1970s saw the police in Strathclyde, Scotland, considering the use of a helicopter and the Met awarding a new contract, which introduced the Enstrom F28 into police operations over London. These involved numerous searches for suspects, escorting VIPs and searching for missing persons, mirroring in many ways the air support roles delivered by NPAS today. Visits by a US president, the Queen's Silver Jubilee and a variety of royal visits across the country saw forces in Hertfordshire, West Yorkshire, Greater Manchester and Nottinghamshire, for example, hiring in helicopters to provide air support cover. In 1977 the police in London became the first civil user of the Marconi heli-tele when it flew over Notting Hill for the carnival, which attracted crowds of around one million. Air support was planned after the event the previous year had erupted into widespread street fighting, the police

BELOW Hughes 269, pictured with police markings and believed to be at Biggin Hill, in 1971. *(Brynn Elliott, Air Support Archive)*

ABOVE Helicopter Hire Alouette G-AWAP fitted with the Marconi Mark 2 heli-tele turret. *(Brynn Elliott, Air Support Archive)*

being caught without sufficient planning and no defensive equipment. Officers cornered by stone- and bottle-throwing crowds found themselves using dustbin lids as defensive shields.

By 1978 a newer Mark 2 version of the Marconi turret, this time a ball-mounted externally fitted camera, had been developed, but few were available for use. New Scotland Yard contracted Marconi to supply them with the coverage and it was Marconi's choice to employ an Aérospatiale Alouette from Helicopter Hire Ltd on the operation. This helicopter was closely allied to Marconi and their heli-tele, and over the years it undertook the majority of the company's airborne development trials and operational

contracts. The Alouette was also used successfully with the Marconi Mark 2 heli-tele in a variety of policing operations, including in April 1979 at a National Front demonstration in Leicester, where the 500-strong demonstrators were policed with help from a helicopter.

In mid-1979 the first UK civil helicopter to be mounted with the Mark 2 heli-tele was a Bölkow Bo 105 G-BFYA, and it undertook coverage of the Notting Hill Carnival in that year. Its design ideals and four-bladed rigid main rotor conferred a legendary robustness, but it was always regarded as an uncomfortable and noisy aircraft. The Bo 105 eventually became central to UK police air support, but at the time there was little interest being shown in twin-engine helicopters, which were expensive but provided good levels of safety.

History was to show that police aviation in the UK was to take a largely rotary path; nonetheless there remained pockets of strong belief in the fixed-wing alternative. The primary advantage of this is financial, with the disadvantage remaining an inability to hover over the target. Many continued to argue and demonstrate that many of the tasks undertaken by a helicopter could be performed equally as well by fixed-wing types. For example, NPAS recently procured four Vulcanair P68 twin-engine fixed-wing aircraft and these will enter service by 2020. The belief is that a future air support service will be provided by a blend of helicopters, fixed-wing aircraft and drones; perhaps the air support pioneers

RIGHT Bölkow Bo 105 G-BFYA carrying the Marconi Mark 2 heli-tele over Notting Hill in 1979. *(Brynn Elliott, Air Support Archive)*

judged the role of helicopters correctly. By the end of the 1970s McAlpine Helicopters Ltd, based in Hayes, was offering a single-engine Aérospatiale AS350 for police operations and talking about the AS355 twin-engine variant. The Metropolitan Police had decided that it needed its own full-time unit and that a twin-engine helicopter was the way forwards.

Growth in police air support: the 1980s

The range of twin-engine helicopter types available in the late 1970s was quite restricted, and the helicopters on offer were medium to large in size. The Met decided to buy the brand new US-manufactured Bell 222 helicopter. It is interesting to note that this decision to buy helicopters might have been a few years premature, as the new Aérospatiale AS355 Ecureuil, the TwinStar, was to come on to the scene marginally too late for consideration by the New Scotland Yard planners. The Twin Squirrel, as it became affectionately known, was to be the police helicopter for many forces.

The history of the Metropolitan Police Service ASU from the arrival of the Bell 222 fleet in 1980 to the move to NPAS in March 2015 is covered in detail in the next chapter. Meanwhile, interest in police air support continued to develop across the rest of the UK, with other forces perhaps feeling that the Met with its Home Office-funded status was receiving a better deal than they were financially. West Mercia Police continued with a hire arrangement for a police helicopter and worked in partnership with Dyfed-Powys Police, while Hampshire along with other forces experimented with a Partenavia P68 fixed wing manufactured in Italy. The P68 Observer offered a complete glass front end and afforded visibility similar to that found in a helicopter. However, its trials in Greater Manchester and West Midlands were curtailed by a crash and it never really broke through in the UK. Some 38 years later the NPAS fixed-wing fleet due to enter service soon is effectively the same P68 aircraft, now manufactured by Vulcanair.

In 1982 a forward looking infrared (FLIR) thermal imaging sensor was used for the first time. Police units had been largely restricted

ABOVE The Bell 222 was able to fly with an early IR sensor. However, it was not externally mounted and had to be internally fitted and pointed out of a window by the air observer. The problem it encountered seeing through the Plexiglas window meant the windows were removed when it was flown. This is one of the Met's helicopters.

to operating in daylight hours and the FLIR was to be revolutionary, as it enabled police to see in the dark: it was effectively the future of police aviation. The FLIR sensor electronically picked out differences in temperature between objects within in its range; although the technology was not new, its application in airborne law enforcement was. One of the major challenges facing the manufacturers was that the FLIR sensor couldn't see through the cockpit's Plexiglas windows and needed to be mounted externally, a separate installation to the daylight heli-tele. The luxury of multi-sensor turrets was many years away. The police in Greater Manchester, Merseyside, Strathclyde and Warwickshire all requested a downlink-capable Marconi turret on aircraft in the Helicopter Hire fleet.

It was no longer considered enough just to have eyes on a target: the control room and senior officers wanted to see pictures from the helicopter day and night. Modern air support was perhaps born in 1982, and this capability is taken for granted across the UK by all 43 forces today. The demand for air support was

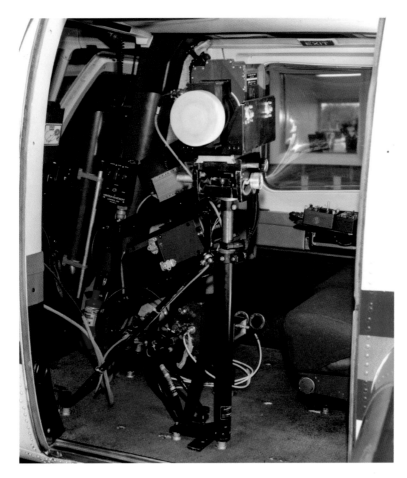

ABOVE The IR sensor here is mounted on a pedestal inside one of the Met's Bell 222 helicopters.

BELOW The Edgley OA7 Optica was widely promoted as a police aircraft and was extensively trialled by Hampshire Constabulary. Sadly the trial ended in tragedy, when a crash of the Hampshire Optica G-KATY resulted in the death of the two police officers who were on board. *(Brynn Elliott, Air Support Archive)*

growing and a number of local arrangements were in place; for example, the Met's Bell 222 helicopters were increasingly travelling across the country to service major incidents and VIP visits. Hampshire Police had for some time been experimenting with the use of fixed-wing aircraft, and when NPAS was just about to come into being Hampshire was one of a small group of forces, along with Cheshire and Greater Manchester Police, who were using these aircraft for air support policing. During 1982 the British police were being targeted by the sales teams behind the British-designed Edgley OA7 Optica project. The Optica was a radical design that promised to provide police and other agencies with a three-seat light observation aircraft with the economics of the familiar Cessna single but visibility for the crew that was equalled only by the best helicopter design. It achieved this by placing the engine and propulsion fan behind the cabin.

The Optica was a design almost wholly reliant on visual observation, and it would not have been able to carry the modern FLIR and heli-tele TV sensors that equip most police aircraft across the world. Hampshire Police decided to test the Optica for the Home Office and the aircraft assigned to the Hampshire trial was registered as G-KATY. However, the trial was to end in tragedy after a crash in May 1985, which claimed the lives of the pilot and observer and entirely destroyed the Optica. After a promising start, the project was confined to the history books.

AS355 in the ascendant: the 1990s

By the start of the 1990s the AS355 was in a great position to be the future police helicopter. A new AS355F2 had been equipped for police service as a demonstrator, being equipped with a more utilitarian interior, a searchlight, a FLIR 2000 sensor pod and sliding doors for photography. The concept of the police role-equipped helicopter was born, and the sales team (led by ex-police officer David Lewis) was chasing the police market with gusto. The helicopters had a red 'jam sandwich' marking with police signs, which made them look like flying police cars. At the same time a rival company, Police Aviation Services based in Gloucester, was also beginning to chase the police market and became one of two primary suppliers. By the beginning of 1990 the purchase of the first AS355s by police forces took place, when West Midlands Police and then the Merseyside Police Authority were impressed by the performance of the helicopter patrols. On delivery, the first helicopter had basic equipment fitted, high skids, a searchlight, a public address system referred to affectionately as sky-shout and an ability to carry a stretcher. West Midlands stated an intention to purchase an infrared camera for night vision use, and in 1990 bought a FLIR 2000 unit, which was fitted under the nose of the helicopter. In April 1991 this was one of the first UK police units

ABOVE The Bo 105 was a popular choice for police air support units, and this one was in service with Wiltshire Constabulary in collaboration with Wiltshire Air Ambulance. The dual emergency services role was popular for economic reasons, but the life-saving role of the helicopter meant shared aircraft were often unavailable for police work. *(Brynn Elliott, Air Support Archive)*

with a microwave downlink capability for the FLIR, a facility that was linked to a Canon video recorder that the helicopter carried.

Other ASUs started to appear across the UK, and the Association of Chief Police Officers (ACPO), the Home Office and the CAA started to look in detail at police aviation. The capital

BELOW The Sussex police and air ambulance MD902 helicopter, which was based at Shoreham Airport. *(Brynn Elliott, Air Support Archive)*

investment in this area was, as it still is today, a major consideration for chief constables and police authorities. The cost of a helicopter made it one of the most expensive capital assets that a police force could seek to procure, and the decision to buy or lease was a major consideration. By the 1990s a number of helicopter variants were present in the UK police market, with the emerging leader being the McAlpine Helicopters-promoted Aérospatiale AS355 Twin Squirrel, which was presented as the perfect twin-engine police helicopter.

By 1993 the Metropolitan Police had purchased the first of its Aérospatiale AS355N Twin Squirrel helicopters. This was followed by a further two AS355N helicopters; by 1996 the fleet of three Bell 222 helicopters had been phased out and replaced by three AS355N

ABOVE The North Wales Police AS355 helicopter with an early Wescam sensor turret fitted under the nose; this was the common mounting place for the AS355 fleet. *(Brynn Elliott, Air Support Archive)*

RIGHT Dyfed-Powys Police entered the air support arena with the AS355 pictured here. Eventually they would replace it with the faster Augusta Westland 109 Power helicopter, which was retired from service upon joining NPAS. *(Brynn Elliott, Air Support Archive)*

LEFT The first helicopter for West Midlands Police was also an AS355, which was based at Birmingham Airport. This was replaced with an EC135, which was subject to a criminal firebomb attack and was written off. The helicopter was eventually replaced with another EC135, which is still in service today with NPAS. *(Brynn Elliott, Air Support Archive)*

RIGHT A very early promotional shot of the Essex Police AS355, which was replaced with an EC135 G-ESEX; both flew out of the recently closed Boreham airfield. G-ESEX is still in service today with NPAS and has been re-registered as G-POLF, now operating out of NPAS London alongside the EC145 fleet. *(Brynn Elliott, Air Support Archive)*

helicopters. The Eurocopter BK117-C1, the predecessor of the BK117-C2 (EC145), was demonstrated to police in the UK at this time, and would eventually be purchased by Devon and Cornwall in 1998. By the late 1990s a new contender for the UK police market had also emerged: the MD Helicopters' MD902 Explorer. This was revolutionary as it didn't have a tail rotor, and as a result was both quieter and considered safer for ad-hoc landings. The MD902 Explorer was eventually to prove a very popular helicopter with several forces including Greater Manchester, Cambridgeshire and South Yorkshire Police operating them when they joined NPAS.

RIGHT The Lancashire Police AS355 helicopter was role fitted in standard police fit with a FLIR Leo 400 camera sensor and an SX16 searchlight. Lancashire was ultimately to trial the use of Airwave radios in police helicopters. *(Brynn Elliott, Air Support Archive)*

ACPO and the Home Office decided that it was necessary to provide some strategic direction for the growing UK police aviation market. Captain Max Kenworthy, a veteran aviator, was brought in as the Home Office police aviation adviser, a role that would continue until the formation of NPAS in 2012. A ten-year strategy was developed, outlining an intention to provide access to police aviation for every force in the UK. Forces were encouraged to invest in the area and the Home Office promised to provide support in the form of capital grants to assist in the purchase of helicopters, role equipment and building bases. Forces could bid annually for a contribution of up to 40% of the capital costs of their air support procurement activities. This proved invaluable, as it meant that forces were able to spread the cost of buying a helicopter over a couple of financial years, the Home Office grant providing a significant boost to spending power. The strategy was working well when the first of the newly launched Eurocopter EC135s entered service with the Central Counties ASU. The EC135 was a modern helicopter with a spacious cabin and a glass cockpit, and was to become the aircraft of choice in almost all future purchases.

A fast-changing policing landscape: the 21st century

Helicopter procurement, supported by the Home Office capital grant, continued throughout the 2000s, with at least two helicopters bought each year. The threat from terrorism changed dramatically on 11 September 2001 with the attack on the World Trade Center in New York, and the police service started to look at how air support could support them in this fight – with the very real fear that an attack could take place in the UK. On 7 July 2005 these fears were realised when terrorists struck in the heart of London, with near simultaneous attacks on the London Underground and in Russell Square. The Met's Twin Squirrel helicopters sprang into action, with two of them providing near seamless coverage of the aftermath of the attacks. The value of air support in the coordination of the emergency services response, together with the gathering of evidence and the provision of excellent situational awareness was realised fully that day, and also in the weeks that followed.

In 2008 a working group was established by Sir Bernard Hogan-Howe, who was then

the chief constable of Merseyside Police and the ACPO lead for police aviation. This working group included representatives of all the air support expertise available around the UK at the time, and sought to determine the strategic direction that air support should follow over the next decade. Across the UK a number of police forces decided it was time to replace their ageing helicopters with a new and more up-to-date variant. The Met's order for three Eurocopter EC145 helicopters was followed

ABOVE When the South East Regional Police Air Support Unit (SERPASU) ended in the early 2000s, the force procured a new EC135 G-SURY. This was fitted with high skids and the 'MacPod', which housed all of the role-equipment electronics beneath the helicopter in a demountable pod. This was very heavy and much maligned by crews.

closely by Devon and Cornwall, who ordered an EC145 to replace their BK117-C1. The Home Office police aviation adviser, by this time Captain Ollie Dismore, achieved a first for

LEFT NPAS inherited many EC135 helicopters that are still in service today, G-EMID was the East Midlands Consortium helicopter and it is seen here at Lippitts Hill sporting new NPAS livery in 2018. Note that the MacPod is still fitted to this helicopter, which is one of the older NPAS airframes.

had an identical role-equipment fit; this was based upon that of the Met's new EC145 fleet. In 2009 ACPO published a report called the 'Police Air Operations Review of the National Strategy 2009', which made a number of recommendations linked to the establishment of a national ASU for England and Wales governed by a single corporate body. The concept of national collaboration for air support was born, and this would eventually lead to NPAS.

NPAS is born: 2010 to date

In early 2009 ACPO, which is now the National Police Chiefs Council, published a ten-year strategy for police aviation. Previous strategies had focused upon the development of air support across the country so that all the 43 police forces in England and Wales could gain access to a helicopter. The new strategy was led by Sir Bernard Hogan-Howe, then the chief constable of Merseyside Police and the national lead for police aviation. It advocated better collaborative working between the various ASUs in an effort to maximise air support capability, saving money through borderless tasking, joint procurement and efficiencies of scale. The recommendations covered such

ABOVE One of the NPAS EC135 fleet G-POLF post-mid-life upgrade. The helicopter was formerly G-ESEX and is now seen fitted with a Wescam MX10 and Nightsun II on the starboard side step bolsters.

UK police aviation by securing the first bulk order of seven identical Eurocopter EC135P2 helicopters. This mini-fleet would save money, as they were ordered at the same time and

RIGHT The South Wales EC135 was based at St Athan pre-NPAS. Uniquely this helicopter was owned and operated by Babcock (formerly Bond Helicopters) on behalf of South Wales Police, with the force providing only the Unit Executive Officer to manage the unit and TFOs. This was one of three EC135s used to service contracts with South Wales and Police Scotland. Upon joining NPAS, the helicopter returned to Babcock. *(Gary Smart)*

subjects as national aircraft, equipment and fuel procurement, national standards for training and operations, insurance and safety equipment. It was felt that the only viable option was for a lead force to take on the running of NPAS, as a bolt-on to their daily policing operations. West Yorkshire Police stepped up to take this role. So it was that the concept of NPAS was born, and between 2010 and 2012 a small team of staff worked up proposals for the bringing together of air support across England and Wales for the first time. The Police Act 1996 provided a collaboration agreement that would outline how the national service would be governed, financed and run. At the time of writing Scotland and Northern Ireland continue their more conventional approach to air support. Police Scotland to this day has a contract with Babcock to deliver helicopters, pilots and maintenance, with the whole of Scotland covered from Glasgow and the use of locally based drones becoming more widespread. The Police Service of Northern Ireland (PSNI) had always been at the forefront of air support use, having first used police helicopters during the height of the Troubles that preceded the Northern Ireland Peace Agreement in 1994. PSNI operates a mixed fleet of fixed-wing aeroplanes and helicopters, and took delivery of their Eurocopter EC145 helicopters in the years after the Metropolitan Police brought theirs into service.

At the peak of air support in England and Wales there were some 33 aircraft, predominantly helicopters, operating from 29 bases and serving most of the 43 forces in England and Wales. The financial pressures on policing, which continue today, perhaps meant that without the formation of NPAS police air support would have slowly reduced as forces sought to balance the books and save money. The Achilles heel of police air support has always been that it is difficult to demonstrate its value in a clear and tangible way, and often those with control of the purse strings have seen air support as a luxury rather than an essential part of modern policing. Today, the need to take many millions of pounds out of the NPAS budget has significantly reduced the number of helicopters flying in England and Wales, but the service continues and is still delivered in a coordinated and borderless fashion to 43 police forces and many partners. At the time of writing a review is under way to determine the direction of travel for air support over the next decade, and a ten-year police aviation strategy based upon a clearly defined user requirement will be published during 2019. Who knows where the next decade will take police air support? Only one thing is clear: it will involve the use of unmanned aviation in the form of small drones and a blended mix of rotary and fixed-wing aircraft.

ABOVE EC135 G-CPSH, formerly operated from RAF Henlow by Hertfordshire and Bedfordshire Police as part of a consortium with Thames Valley Police. The Henlow unit was closed upon joining NPAS in 2012.

Chapter Two

The Metropolitan Police Air Support Unit

From a base at Lippitts Hill overlooking north London that was first used in 1967, air support is provided to the capital. Permanent air support started in 1980, and since then three different helicopter types have been maintained and operated by the teams of engineers, pilots, tactical flight officers and control room staff to police London from the air.

OPPOSITE A Metropolitan Police EC145 pictured over London. The photograph was taken from inside an Arena Aviation helicopter used for aerial filming during the Red Bull air race in 2007. *(Arena Aviation)*

Permanent air support arrives in London

The Metropolitan Police was one of the early adopters of air support, having dabbled with aviation since the 1920s. By the late 1970s a decision was taken that the Met should have permanent and ready access to air support. The extensive efforts of some of the Met's best crime-fighting police officers during the 1970s demonstrated the clear benefits of a police helicopter in the fight against crime. They chalked up numerous successes as they used the benefits of an aerial point of view to search for, follow and ultimately bring into custody armed robbers, burglars and other criminals.

The benefits of police aviation were seen very differently by the Met's senior officers, who valued the aerial observation platform as an essential tool in the policing of public order events both in London and across the UK. The ability to deliver live imagery from the helicopter via a downlink had already been proven and the benefits of the aerial overview provided to the command and control teams policing major events were immense. It is fair to say that the permanent Metropolitan

Police ASU was founded as a public order resource, and as Sergeant Tony Mepham, one of the first permanent air support officers explained, 'public order paid the bills and as such it was always a priority'. The concept of air support was championed by Central Command Complex (CCC) based at Scotland Yard, who saw the benefit of aerial imagery for managing public order and other significant policing events. The transport department was commissioned to provide the helicopters and the Met's technology department, based at Newlands Park near Sydenham, provided the technology. This cooperation between different departments for the delivery of air support was to remain for the first few years, but it was not without its challenges.

In the late 1970s the transport department decided to buy three helicopters, which would be based at a brand new facility built at the Met's Lippitts Hill site in the heart of Epping Forest. There were a number of helicopter manufacturers who were competing to get into the emerging police market; however, one, which would become the market leader, was just a little too late to attract the Met's attention and secure its custom. As previously mentioned,

BELOW The Met was a launch customer for the new Bell 222, and trials took place in Fort Worth, Texas, for the mounting of the Marconi heli-tele turret. Initial plans to mount it on the port side (seen here in the photos) were shelved as it caused aerodynamic problems. The Bell 222 was ultimately delivered with the turret mounted on the starboard (right) side.

the Aérospatiale AS355 Ecureuil Twin Squirrel helicopter with its comfortable ride, powerful twin engines and relatively low running costs was to arrive on the scene just too late: when David Lewis of McAlpine Helicopters took his police variant demonstrator around the UK over the next decade, he sold the concept of the role-equipped helicopter to many police forces. The Met made a decision to buy an emerging new helicopter type from US-based Bell helicopters, and the order for three Bell 222s was placed.

This helicopter was brand new, and came equipped to a very high level of technological capability. Its Decca Tans navigation system allowed the programming of ten waypoints, and was the envy of many aviators who longed for the sophistication of an automated navigation system. Leather seats adorned the cabin, and it was painted in the Met's familiar vehicle livery of white with the 'jam sandwich' stripe, the Metropolitan Police crest adorning the side panels and tail boom. After testing in Fort Worth, Texas, to establish how best to accommodate the Marconi heli-tele turret, the first of the Bell 222s arrived at Lippitts Hill and entered service. The Met's ASU was born. In these early days the only police involvement was the posting of three sergeants, including Sergeant Tony Mepham,

and 16 police constables, selected according to what they had achieved as operational officers and also who they knew. Many were traffic officers who it was thought would have good navigational skills.

In the early 1980s a brand new hangar was built to accommodate the ASU and its officers and engineers. The crews worked two shifts of 7am–3pm and 3pm–11pm Monday to Friday, with weekends reserved for pre-planned public order events. The unit operated in a way that would be the envy of present-day crews: they patrolled proactively, with four-and-a-half flying hours, spread over three flights allocated per shift, every day. These patrols almost took the form of scheduled flights, with one taking off to monitor traffic at 7.45am and further flights following at key times throughout the day. The air observers, an early version of today's TFOs, were not permanently attached to air support: they worked in teams of four, with each team staffing the two shifts for a week before returning to their normal policing duties. With minimal specialist training, a lack of aptitude for the role among some of them, and the lack of experience and continuity caused by only flying one week in four, the success of the unit was variable.

The role of the sergeant was to manage the

observers. Every time the helicopter was flying, a sergeant was required to be in the air support control room to monitor the flight for safety. The tasking of the helicopter was still done by CCC from Scotland Yard, and for many sergeants the burden of having to sit, often doing nothing, in a control room at Lippitts Hill was tiresome. The Met's 222 fleet was by now routinely policing major events and VIP visits across the UK and was one of a small number of police helicopters delivering air support to forces. The Bell was comfortable to fly in, but its performance was lacking and it just wasn't up to the role. Stories are told of loading the Bell 222 up with equipment and fuel, but no crew, and taking off from Lippitts Hill through a gap in the trees hoping that it would gain height and keep flying. The TFOs meanwhile had to travel by road to the nearby Stapleford Aerodrome to be picked up by the helicopter, which could then take off easily from a runway.

By the mid-1980s Tony Mepham and his team were delivering professionalised air support and the police officers were beginning to gain some control and influence over the way it operated and the technology it carried. Tony established a small team of six permanent air observers, who were selected for the role owing to their air support ability. The helicopters were flying nine hours a day across the fleet and were cared for by a team of engineers at Lippitts Hill who maintained, cleaned, refuelled and marshalled the helicopters in and out of the site. The sight of engineers wearing pristine white overalls and crawling over the helicopters in maintenance was a welcome and reassuring sight. It is clear that bringing the Lippitts Hill teams together for one common purpose was to prove one of the biggest developments of the period. The helicopters were routinely

RIGHT The Bell 222 is marshalled out of the north hangar at Lippitts Hill by one of the Met's engineering staff. *(Brynn Elliott, Air Support Archive)*

fitted with a large Marconi daylight TV camera turret mounted on the starboard stub wing strut, early manufacturer trials having found that a port side mounting caused aerodynamic issues. One of the major innovations was the introduction of the very first infrared (IR) sensor that was manufactured by Thorn EMI; this was mounted inside on a pedestal. Problems were encountered as the IR could not see through the Perspex windows, and as such

BELOW Met engineers in their pristine white overalls complete maintenance inspections on two of their Bell 222 fleet. *(Brynn Elliott, Air Support Archive)*

31

ABOVE The Bell 222 pictured at Lippitts Hill with three of the crew from the late 1980s. PC Terry White on the left in the photo retired as a sergeant from the air support unit in 2004 having served on air support for over two decades. Terry was one of a very small number of air observers who flew in every one of the Met's three airframe types (Bell 222, AS355N and EC145).

BELOW Met Bell 222 G-METC is pictured under maintenance in the north engineering hanger at Lippitts Hill.

ABOVE When the Met ASU started in 1980 the proud ASU staff developed a crest for the unit and a clever motto 'Above All We Serve'. A very small number of mounted versions of the crest were produced before a senior officer reprimanded the unit for the unauthorised use of the royal crown. One of the few surviving mounted crests is shown here. The motto continued throughout the years to come for the MPSASU.

LEFT Met Bell 222 G-METB looking broken and muddied after a crash landing at Hatfield in 1984. Thankfully all of the crew and passengers escaped with no loss of life; some were injured, though.

IR operations were usually conducted with the helicopter door open; this could prove very cold in winter.

Challenges and setbacks

The Met's aviation history is not without its setbacks. In 1984 the second of its Bell 222 helicopters (G-METB), with two passengers in addition to the crew of three, crashed upon attempting a practice instrument approach and landing at the British Aerospace factory in Hatfield. As the final approach was made the helicopter appeared to have lost yaw control, and although the tail rotor seemed to be rotating normally the pilot had no tail rotor authority. A full emergency was declared and the pilot skilfully managed to control the aircraft as it attempted an emergency landing. It was only once the helicopter was on the runway that there appeared to be any ability to counter the torque from the main rotor using the brakes, and the end result was a dramatic-looking crash that ended up with the Bell 222 upside down, battered and covered in mud beside the runway. Everyone on board escaped, although they were all injured to one degree or another; thankfully none of the injuries were serious. The sight of a battered and muddied Metropolitan Police helicopter was a sobering one, and this reminded the Met and other police aviators of the potential hazards of police aviation.

The Met ASU bounced back from this incident and continued to fly the remaining two Bell 222

helicopters in its day-to-day patrols, as well as playing a public order policing role and catching criminals. In 1987, however, there was another setback when the first of the Bell 222s (G-META) suffered an engine explosion upon landing on the Lippitts Hill helipad. Thankfully the explosion was largely contained within the engine cowlings, and as it occurred on touchdown the helicopter landed safely and no one was hurt. It transpired that the No 2 engine had not been performing as

BELOW A closer look at the port side engine cowlings shows the extent of the damage caused by the exploding engine. There is no doubt that the timing and location of the explosion saved lives.

well as expected during the latter part of the flight, and the pilot had elected to return to Lippitts Hill as a precautionary measure. There were two passengers on board wearing headsets, and the increased cabin noise from the engine had been attributed to their live microphones. G-META did not look badly damaged; indeed it appeared to be neatly parked on the helipad. However, the damage to the cowlings and engine bay was expensive. This was another blow to the Met's air support ambitions.

Everything changes

In 1994 things were to change when the CAA, in response to an ever-growing police aviation market, issued Civil Aviation Publication 612 (CAP 612), the Police Air Operations Manual (PAOM). This set out everything that a police ASU must do to comply with the regulations and be issued with a Police Air Operations Certificate. When Inspector Phil Whitelaw arrived at the ASU he described this as a game changer. Suddenly the police were in control, with a police officer taking on the role of Unit Executive Officer (UEO), which Phil undertook for the Met, and being accountable for compliance with the requirements of the PAOM. This effectively professionalised police aviation, and every force wanting to establish its own ASU had to invest, in aircraft and personnel terms, in a way that complied with the requirements of the PAOM (which was referred to as the POM). The PAOM outlined helicopter performance requirements, and the Met found that its Bell 222s were unable to meet these. A dispensation was obtained to allow operations to continue while a replacement fleet was sourced, but the Bell 222s' days were numbered.

By the mid-1990s the Met had introduced three AS355N Twin Squirrel helicopters. The first was G-METD, which arrived fitted with a Star Safire thermal image camera and was soon to be upgraded to the new FLIR Leo 400 dual TV and IR turret, which was to serve the Met well until 2003. A partnership was formed with Surrey Police, and this saw the opening of a satellite base at Fairoaks Airport in Chobham, near Woking. One of the Met's three AS355N helicopters was based there, along with a crew of TFOs drawn from the two forces. This partnership involved Surrey paying for 300 flying hours annually on top of the 3,000 hours paid for by the Met. The ASU was renamed the South East Regional Police Air Support Unit (SERPASU) and the three helicopters were registered as G-SEPA (formerly G-METD), G-SEPB and G-SEPC. The partnership with Surrey continued for many years, but it was not without its challenges. The Met undoubtedly enjoyed a faster service to the south-west of its area, the opposite side of London to Lippitts Hill, which was to the north-east; however, the draw of the metropolis and its high levels of risk and demand for air support meant that Surrey often came a poor second in the tasking process, and did not receive the cover it deserved. Early in the next decade plans were under way for Surrey to

BELOW The first of the Met's new AS355N Twin Squirrel helicopters is fitted with police role equipment before delivery to Lippitts Hill.

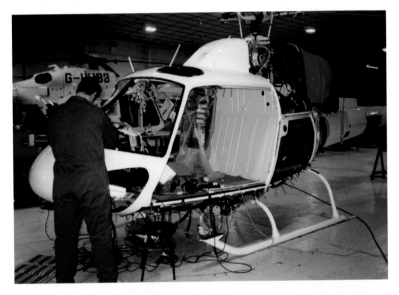

LEFT The first of the Met's new AS355 helicopters had the registration G-METD, as it followed on from the three Bell 222 registrations. This was later to be changed to G-SEPA when the collaboration with Surrey Police commenced. It is pictured here at Heathrow Airport, with the Star Safire turret fitted; this provided only thermal image capability. This was later replaced by a new FLIR Leo 400 dual sensor turret, which provided both daylight and infrared imagery.

LEFT G-SEPA, a Met Police AS355N Twin Squirrel, is lifted on to a low loader as it is sent away to be repainted. It was to return in a smart new blue and yellow colour scheme some weeks later.

LEFT G-SEPA is pictured in flight over north London after returning with its new paint job. The airframe had completed 10,000 flying hours and it was decided it was due for a respray.

ABOVE One of the Met's three AS355N helicopters, G-SEPC, in flight near the ASU base in north London. The helicopter had been upgraded to include a new dual sensor FLIR Leo 400 turret, which now provided daylight and infrared imagery.

break away from the partnership with the Met and buy its own Eurocopter EC135 helicopter (G-SURY). By 2003 the Met was again operating its three helicopters from Lippitts Hill in Essex.

By 2002 the Metropolitan Police Authority (MPA) was formed. This meant that for the first time the Met could bid for capital funding. An upgrade was urgently needed as the Met's fleet of three AS355N Twin Squirrel helicopters was getting old, and was lagging behind the rest of

LEFT The arrival of the EC145 fleet saw another redesign of the Met's ASU logo, and a new badge was produced; this adorned the crews' flight suits with pride for the next eight years. These local badges were replaced with a corporate NPAS badge in 2015 to bring together a sense of national identity.

LEFT The use of aviation by the emergency services was expanding quickly, and in an effort to explore use by the Fire Brigade a trial was undertaken in London, with a BK117-C1 providing the platform. This publicity photo shows the *Daily Express*-sponsored London Air Ambulance alongside the Met's AS355N and the Fire Brigade trial helicopter over London. All three emergency services served London from the air. The fire trial was not continued for a number of reasons, but it has been revisited since. However, the complexities of landing in London rule out a lot of potential fire brigade uses. As an aside, the London Air Ambulance is now (2019) sponsored by Virgin.

UK policing in terms of its equipment. The crunch came with an operational requirement to read a vehicle number plate from a distance without being detectable on the ground. The FLIR Leo 400 turret, which had served the MPS well, was simply not up to the job, and the technological advances in sensor technology had been so great since its purchase that it looked decidedly poor when compared with modern sensors. A mid-life upgrade programme commenced, and this saw the three helicopters fitted with a Wescam MX15 sensor, the very best on the market, interfaced with a Skyforce Observer moving map system, Airwave digital radios and a digital video recorder. The AS355N is a fantastic helicopter and very capable in its police role, so the Met with its upgraded role equipment was once again delivering the very best air support: it had caught up with the other police forces, which were largely enjoying their modern Eurocopter EC135s or MD Helicopters' MD902s and updated equipment.

The golden years of Metropolitan Police aviation

Every member of staff who has contributed over the years to the development of air support in London can look back with pride at their achievements and the service they have given. The reassuring sound of a helicopter overhead at an incident is one that every police officer recognises, and 'India 99' is a call sign that will always represent a unique and much-

ABOVE G-SEPC on a rare trip out of the capital, flying to Wales for flight trials. This photo shows the author taking photographs through the open rear door during the trials.

loved tactical option. Thankfully, the work of NPAS means that this reassuring presence continues today. For many staff the golden era of Metropolitan Police aviation began in the 2000s, when the unit expanded from a 16-hour coverage period to 24-hour policing. This move happened largely by accident after a discussion over a few pints with the chief superintendent who was responsible for the ASU, Simon Humphrey. During a meeting in a pub around the corner from New Scotland Yard with the three ASU sergeants

BELOW The Met's AS355N, G-SEPC, pictured in flight with the newly installed Wescam MX15 turret fitted under the nose. This sensor would later be transferred to the new EC145 fleet and remains in use today.

(the author Richard Brandon, John Gleeson and Andy Hutchinson), the conversation turned to the fact that policing was a 24/7 business and that air support should be the same. Chief Superintendent Humphrey simply asked 'Why aren't you 24 hours then?' and the answer from the three sergeants was 'I don't know, boss!' The direction came to 'make it happen', and they left the pub a few drinks later drawing lots as to who was going to break the news to Inspector Phil Whitelaw, the unit's boss. The unit went 24 hours a few months later, with the pilots and TFOs working a 12-hour shift pattern.

Met ASU establishment

At this time the ASU had some 50 members of staff working round the clock to deliver an operational service to London that guaranteed 24/7 coverage, flying an average of eight hours a day and undertaking 24 to 30 operational tasks each and every 24 hours.

The Metropolitan Police ASU team comprised:

Police officers
- Police Inspector – UEO and Accountable Manager
- Police Sergeant – Deputy UEO and TFO
- Police Sergeant – Training Manager and TFO (also qualified as UEO)
- Police Sergeant – Operations Manager and TFO (also qualified as UEO)
- 18 × Police Constables – TFO

Pilots
- Chief Pilot
- Deputy Chief Pilot
- Flight Safety Officer
- 8 × Line Pilots

Engineering staff
- Chief Engineer – B1 licensed and type rated
- Deputy Chief Engineer – B1 licensed and type rated
- 3 × Licensed Engineers – B1 licensed and type rated
- Avionics Engineer – B2 licensed and type rated
- 2 × Engineering Fitters
- Quality and Compliance Manager – B1 licensed engineer
- Engineering Stores Officer
- Technical Records Officer

Control Room staff
- 5 × Police Staff Controllers
- Intelligence Analyst.

BELOW A Met ASU photo taken in May 2013 upon the retirement of Sergeant Terry White, who is pictured in the front row, fifth from the right and next to the author. The porcelain pig in front of the group featured widely in unit photos and became something of a mascot; however, it was never clear where this originated: perhaps it's better not to ask! The photo features crews, pilots, engineers and communications officers, and is representative of the Met ASU team.

LEFT The new EC145 helicopters pictured at Lippitts Hill after they entered operational service in July 2007.

The staff at Lippitts Hill were a well-oiled machine, with few members of staff choosing to leave thanks to the high levels of job satisfaction and the family feel at the base. Most departures were police constables seeking promotion, and to this day ex-members of staff are always a key feature at any ASU get-together. The history of the ASU is retained by the storytelling and reminiscing of old and long since retired members of the unit.

Achievements

In July 2007 the Met's final fleet of helicopters, three Eurocopter EC145 helicopters, entered operational service, and the unit started to look forward to the new and exciting roles that the new helicopter enabled. The unit was operating at full strength, with its annual budget of around £6m enabling 3,300 flying hours annually and guaranteeing the availability of a single helicopter 24 hours a day every day. The engineering team worked between 0700hrs and 2100hrs daily with on-call cover out of hours at weekends, and the rate of flying on each airframe was carefully controlled in liaison with the chief engineer to ensure that one helicopter was always in deep maintenance with one operationally available all the time and a second as much as possible. The TFOs and pilots were organised to deliver 24/7 cover through two 12-hour shifts with a middle shift (typically noon to midnight) being available whenever there was a second helicopter to fly. The rate of flying was an average of nine hours a day, but this varied seasonally – with more flying in the summer

months when the weather was better and less in the winter because fog or poor weather stopped flying. The day-to-day monitoring and an ability to move and flex the engineering schedule to accommodate higher or lower flying rates were crucial to the success of the unit.

The Met was one of only two ASUs in England and Wales to have its own in-house engineering capability. The other was the Essex unit, which established Eastern Counties Police Maintenance based at RAF Wattisham and delivered maintenance to a number of EC135 helicopters. Having a team of engineers with no corporate profit requirements meant it was possible to vary maintenance plans quickly and easily and to achieve excellent service availability. The engineering team at Lippitts

BELOW A proud day for the Met ASU was celebrated when the new EC145 fleet entered operational service in July 2007. The (then) unit boss, Insp Phil Whitelaw (Retd) is pictured (front row, second right) with Chief Pilot (at the time) Capt Brian Baldwin seated to his left, with the author on his right. Sgt John Gleeson (Retd) makes up the front row as the whole ASU team welcomed in the new fleet.

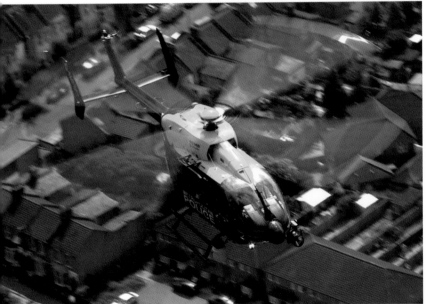

Hill worked with exceptional pride, and made sure that the Met always had the helicopters it needed to meet operational demands. Weekend public order demonstrations required two helicopters and plans were always in place to ensure that this was achieved with sufficient flying hours available on each to meet the demand. Larger events, such as Notting Hill Carnival, which takes place in London over August Bank Holiday weekend, required all three helicopters to be available, a real challenge. The engineering plan was always drafted to build in a split between the three helicopters, ensuring that they did not require base maintenance at the same time – and indeed flying was carefully managed throughout July and August to ensure that the third helicopter (used as a spare) was available to

meet the Notting Hill demand.

Like all ASUs and the NPAS crews flying today, the Met's ASU achieved many things over its years of operation: every year it flew between 2,700 and 3,300 hours and attended around 7,500–8,000 tasks in the Metropolitan Police Area. Around 150 sets of aerial photographs were provided for planning and evidential purposes, many hundreds of suspects were caught and missing or vulnerable persons were located from the air. Some days, however, will always stand out, remaining the subject of many a tale recounted at reunion events.

Terrorist attacks

The Met flew through years of terrorist activity by the IRA, attending many terrorist atrocities such as the Hyde Park bombing in 1982, the Harrods bombing in 1983 and the Brighton bombing of the Conservative Party conference in 1984. The service provided valuable aerial imagery and resource coordination. July 2005 saw an unprecedented attack on London when several devices were detonated by Islamic terrorists on the London Underground and a bus in Russell Square. The Met ASU in a Twin Squirrel helicopter were on scene within minutes of the Russell Square bomb exploding and assisted to coordinate the emergency response to the atrocity while filming and downlinking live video imagery to the command teams of multiple emergency services, which proved to be of vital operational and evidential value. The challenges faced on 7 July 2005 were complicated further by the fact that the Met ASU had dispatched a team of pilots and TFOs to support the Scottish Police in the policing of the G8 summit that was taking place at Gleneagles in Scotland. In the days that followed the bombings the somewhat depleted and stretched Met ASU crews provided continuous air coverage over London, with crews piggybacking each other in two helicopters and flying dozens of hours as police activities and reassurance patrols continued. This rate of flying was only curtailed after it became clear that if the unit did not decrease the rate of flying all three helicopters would run out of hours and end up in maintenance together. A decision to return to normal flying

was taken, and this proved a good one – as two weeks later on 21 July 2005 a failed terrorist bomb attempt led to a manhunt across London that lasted several days and in which the ASU played a significant role. Thankfully, attacks like this are not commonplace, and it is because of their shocking impact that they tend to stick in the memory of the crews involved. In recent years NPAS crews from across the country have responded to terrorist attacks in London and Manchester, with the air support ability to capture key evidence and provide command and control imagery being considered a key operational user requirement.

Royal weddings and state occasions

On a more pleasant note, the Met's ASU provided air support coverage for the wedding of the Duke and Duchess of Cambridge in April 2011. Planning imagery was provided along with a significant programme of security flights and aerial open area and rooftop searches in the days leading up to the event. On the day itself the ASU crews provided live imagery from high above the wedding footprint and also assisted the response to suspect vehicles on the route by providing live downlinked imagery to the Special Operations Room (SOR), call sign 'GT'. Involvement in large-scale state events such as the royal wedding or state visits have always been handled with pride by the ASU crews, who know that their performance and contribution to the policing operation is in the spotlight. ASU involvement runs for months in advance, as planning meetings are attended, aerial imagery is provided, airspace restrictions are drafted and policed, and the ASU role on the day is fine tuned. Large events such as this will always have an ASU sergeant based in SOR, to provide tactical advice to the command team and ensure that the tasking of the ASU helicopters is as required. The need to manage crew hours, fuel stops and helicopter handovers took careful planning, and this was achieved by the ASU liaison sergeant in conjunction with the control room and crew at Lippitts Hill. It was often commented that failure to turn up at or deliver a result at a garden search would

LEFT Working to support a safe and effective visit of a US president has always been a challenge of planning and logistics. Here the EC145 is pictured on the ground at an airport awaiting the arrival of Air Force One, which is seen in the background. The helicopter would soon take off as part of the convoy of US helicopters of US Marine Corps' VIP transport squadron HMX-1.

that was hosting the presidential party. The details of each visit varied, but the role of the ASU crew was always the same: searching and securing the various landing sites, escorting the presidential convoy by land and air, providing additional security measures and above all reassuring the policing command team that the convoy was safe and on time. The planning process involved engagement with the airspace and airport authorities along with US Secret Service, US Marine Corps (HMX-1) squadron personnel, UK government and the policing team responsible for the visit. The highlight was always flying in two Met helicopters in convoy at relatively low level over London with the US presidential convoy, one leading and one at the rear. The sight must have been spectacular for those observing from the ground, while the thrill of taking part was always enjoyed by the crews, who relished the opportunity to participate in a small piece of history.

seldom result in criticism, whereas failure to deliver at a pre-planned state event almost certainly would: failure was not an option.

US presidential visits

F ew state events are as challenging as a visit to the UK by a serving US president: visits by President Bush, President Obama and most recently President Trump have tested the ASU planning and operational capability to the limit. The ASU was involved in the security and escort of the presidential party from arrival aboard the iconic and impressive Airforce One, through the transfer to Marine One, a US Marine Corps helicopter, to landing at the venue

BELOW Air support plays a pivotal role in many annual events that attract large crowds. In this image taken from the infrared sensor, crew members have a unique view of the annual fireworks display on Blackheath Common in south London.

London riots, 2011

T he summer of 2011 was a hot one for many reasons, and when on 6 August a protest march into the police shooting of Mark Duggan in Tottenham turned violent, the police had no idea how far this would spread. The next six days saw some of the worst violence and public disorder that the UK had seen for decades: it was as though the capital was a tinder box, and one small spark in Tottenham managed to ignite the flames across London and indeed other parts of England. The ASU has always played a pivotal role in any public order policing event, and this was no different. The crews from Lippitts Hill scrambled to crew two operational EC145 helicopters, call signs 'India 99' and 'India 98', and took off to

provide live downlinked imagery into the SOR and other local control rooms. The crews were used to seeing violence and disorder on the streets of London, with events such as Notting Hill Carnival, the Million Mask March and other protest marches occasionally witnessing pockets of violence and disorder; but nothing could have prepared them for the burning of buildings and vehicles and the looting of retail outlets in multiple locations all over London. Copycat protests spread across the UK and the ASUs policing various major cities were called into action.

In London the ASU realised that it could not cope: the footprint of the disorder and its intensity were such that the demands for air support in different geographic locations as far apart as Enfield and Croydon, meant the two Met helicopters were quickly operating at capacity. Air support from Essex, Surrey and Thames Valley forces was requested, and on occasion for the first time there were multiple helicopters policing London simultaneously – coordinated by an ASU sergeant as liaison officer in SOR. The role of the helicopter crews was to provide an officer safety overlay, detailed situational awareness and evidence capture. At times they were simply resigned to watching criminality unfold in front of them, with the police having neither the ability nor the capacity to deal with it; this was extremely frustrating. On one occasion, in Enfield, the crew of a Met EC145 were observing as a serial of police officers (a group of one sergeant and six constables) were

separated from their colleagues and appeared to be at significant risk of violent attack. Fearing a repeat of the horrific murder of PC Keith Blakelock in 1985, the crew took the decision to fly low over the rioters, pushing them back as the downwash from the helicopter blew flaming debris towards them. Exactly how low the helicopter flew was never really discussed and the lawfulness of the action was briefly questioned, but there is no doubt that this action saved the lives of those officers, who were able to regroup and regain control. There is no doubt that the role played by police air support in quelling the riots was significant and very welcome.

London Olympic Games, 2012

The hosting of an Olympic Games is a special thing for any country, and in 2012 the honour was to pass to the UK. In July 2005 London was successful in its bid to host the 2012 Games, something that was overshadowed by the London terrorist bombings two days later. The combination of the two events kicked off discussions as to how an ASU could contribute towards the policing of the event some seven years later. The project to replace the ageing AS355N fleet of helicopters was given a boost and the EC145 was selected as the helicopter of choice. The thinking behind the choice of a bigger helicopter airframe was the ability to have new and previously untried

ABOVE Public order policing frequently requires continuous air support coverage, and it is common for the helicopter to land for a refuel at London's heliport in Battersea. EC145 G-MPSB is pictured here during such a refuel, with the visit to the heliport giving the crew a taste of the VIP lifestyle … almost!

ABOVE A rare photograph of all three Met EC145s in flight courtesy of a media helicopter, with which the ASU was liaising during the London Olympics of 2012. The opportunity to get all three flying together just before the Olympics began was too good to resist.

tactical options. The carriage of up to six fully equipped and armed specialist firearms officers (SFOs) and deployment to the ground by fast rope were considered essential operational requirements. The EC145 was capable of delivering these roles in principle, the main challenge being that they had never been undertaken by civilian police helicopters before: they were previously military-only tactics.

Planning, training and preparation for the Olympics continued: the event was set to be the biggest ever policing event for the Met. The eyes of the world would be on London in the summer of 2012, and failing to deliver a safe and successful games was not an option. The EC145 helicopters entered service in July 2007, and with five years to go plans were laid out to enhance the ASU: these included additional aircraft, crews, resources and equipment. Sadly, there was no goose to lay a golden egg, and despite the ASU's optimistic outlook the message came through in 2010 that there was no new money and air support for the Olympics would need

CENTRE The best seat in the house. The helicopter crew had a unique view of the opening and closing ceremonies at the Olympics and many of the other key events as they flew overhead providing live imagery to police commanders and control rooms. The opening ceremony is pictured here. *(Hugh Dalton)*

LEFT The Olympics hockey arena pictured from a Met Police EC145.

RIGHT The EC145 was fitted with an ECMS rope down beam, which enabled the development and delivery of specialist air support tactics from police helicopters for the first time in the UK. The hoist camera developed by Skyquest is also visible: this provided an excellent view of the rope and ground below on the pilot's display.

to be delivered from within existing resources and budgets. Challenges like this were not uncommon, but the ASU engineering staff rose to the challenge, describing the Olympics as similar to a Notting Hill weekend but just longer: three weeks rather than three days. The operational requirement was for three helicopters to be available for this three-week period.

Two helicopters were to be available 24/7 in the three-week period leading up to the opening ceremony, one for business as usual policing and one for Olympic security and other tasking. During the games a third helicopter was to be available every day for a defined period of time (12 to 18 hours), equipped for transportation and deployment of counter-terrorism SFOs (CTSFOs) by air to wherever they needed to be within the Olympic footprint. The amount of daily flying was largely predictable, and on-site engineering support was made available throughout the Olympic period to rectify any unscheduled maintenance issues. The manufacturer of the helicopters to be used, Eurocopter, made spare component parts immediately available, some being included as part of the service by the hour contract and others provided just in case. The effort behind the scenes lasted for many months and included all the ASU staff, equipment suppliers and partners to ensure that every eventuality was considered and nothing could go wrong. The delivery of three serviceable helicopters by a unit that usually had one helicopter of three in deep maintenance, in other words in a million bits in the engineering hangar, was a real challenge of planning and

RIGHT A training sortie in which officers are being roped on to the rooftop of a disused building.

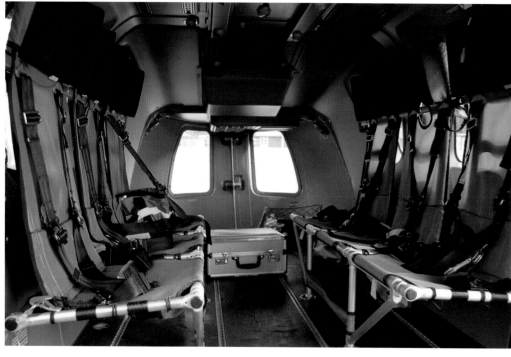

RIGHT The Met's EC145s were designed to be re-roled, and the clear rear cabin allowed for utility seating to be fitted and up to six fully equipped specialist firearms officers (SFOs) to be carried. Pictured here is the empty rear EC145 cabin with canvas utility seating.

BELOW Specialist tactical options such as fast rope required TFOs in fall arrest harnesses to operate outside the helicopter to extend the beam and marshal the helicopter on to target. Note in this picture that the Wescam MX15 sensor has been removed: all police role equipment was demountable to save weight for special roles sorties.

engineering dedication and effort. The ASU had to slow down the daily rate of flying and carefully manage flying hours in the run-up to the summer of 2012. The Met's in-house engineering team worked round the clock to complete major maintenance, thereby ensuring that in the week before the opening ceremony on 27 July they handed over three serviceable helicopters with sufficient flying hours capacity to meet the operational requirement. Plans were also in place to complete the maintenance required during the Games, the 50- and 100-hour inspections, with the minimum of operational impact.

Other Olympic resourcing for the ASU was relatively simple: no leave was allowed and all the pilots and TFOs actively sought to play their role in the Olympics; to be able to say 'I was there, and I policed that event.' History shows us how successful the 2012 Olympics were for the UK: they passed off without any incident and were widely considered to be among the most spectacular ever. The ASU crews and staff, flying helicopters India 99, India 98 and India 97, shared in the joy of this Olympic success and captured some spectacular imagery that showed the transformation of Stratford, east London, from a derelict and depressing wasteland to a spectacular Olympic venue. The view of events such as the opening and closing ceremonies was breathtaking from 1,500ft in a police helicopter. Close liaison with media companies and ATC enabled the police helicopters to safely share the airspace with a multitude of television and other Olympics-related air traffic.

The Met's ASU has been privileged to participate in almost every major event that has taken place in London over the last four decades. The memories are the subject of many a get-together, where crews reminisce about the old days and the experiences they shared. Events such as the Tour de France, Red Bull Air Race, the Queen's Diamond Jubilee and many state visits have filled the Met archive with unique aerial imagery that has been captured from the best seat in the house. The Olympic Games, though, was perhaps the pinnacle of many a Met officer's career; certainly the ASU contribution was at a level that is unlikely ever to be repeated. NPAS now has the privilege of providing air support for future events in London and across the country, and the same crews wearing a different badge continue to rise to each challenge and meet expectations every time.

Transition into NPAS

While the south-east region of England was chosen to be the first to join NPAS, the Met ASU was to be the last to join – allowing the plans for the Olympics to continue without impediment or last-minute change. In September 2012, after the Olympics closing ceremony, the first NPAS region went live. The evolution of NPAS as a whole was lengthy, complicated and at times problematic; it could easily be the subject of an entire book. Suffice it to say that the Met's journey into NPAS was not without controversy, as many of the benefits of air support collaboration (economies of scale, engineering, procurement, etc.) were already being enjoyed by the Met, owing to its size and scale of operation and the fact that it already had its own in-house engineering and command and control capabilities. Finally, in early 2015, it was decided that on 30 March the Met ASU would become part of NPAS, thereby delivering an almost complete national air support service. Only Humberside remained outside the new organisation, its entry delayed by existing contractual arrangements for helicopter purchase.

The next chapter of police aviation in London began with the formation of NPAS London; however, the pilots and TFOs were the same, the EC145 helicopters were the same and for the next few years the ASU command and control facility was the same. The casualty of the change was the in-house engineering department, as the engineering function of NPAS was to be undertaken through a new national contract, which was awarded to Airbus

Helicopters Limited in Oxford. The transition did not really change air support provision, as the same people delivering the same service in the same way meant the overall situation was largely unchanged. The base at Lippitts Hill remains in use today, hosting NPAS London, recently merged with NPAS Boreham, and will continue to provide an air support service across London and the south-east of England, 24 hours a day, until the unit moves to a new purpose-built base at North Weald Airfield in late 2019. The closure of Lippitts Hill and its likely disposal will bring to an end a memorable association with the site and the local Epping Forest community stretching back to the 1960s.

Lippitts Hill

The story of Lippitts Hill begins long before it was given to the Met by the Ministry of Defence in the 1960s. When the Met was looking for a home for its future ASU a number of potential sites were considered, but after viewing the range of facilities available at Lippitts Hill Camp, including the promise of a well-equipped workshop, the choice was obvious. Lippitts Hill sits 300ft above sea level and looks down the Lee Valley into London: the view into the city from this former military site is breathtaking from the helipad.

ABOVE The EC145 was also capable of lift and shift of other specialist resources, and routine training with police dogs was commonplace at Lippitts Hill. In this picture a dog handler can be seen carrying the dog from the helicopter, to ensure safety is maintained at all times. This is a far cry from the deployment of police dogs from a crate outside the helicopter in a low hover, which was trialled during the development of air support in the 1960s.

LEFT The Lippitts Hill site long before the arrival of air support, in around 1960. *(Brynn Elliott, Air Support Archive)*

CENTRE The three Met EC145 helicopters pictures on the pad at Lippitts Hill just before departing for a rare media opportunity relating to the Olympic Games, when all three Met helicopters were pictured together in flight.

During the war British and American anti-aircraft artillery (AAA) were based at Lippitts Hill to guard the eastern approaches to London. The majority of the concrete structures that remain on site today were erected by the US forces: gun emplacements, sunken block houses and a massive underground control room were supported by permanent roadways and a range of wooden and concrete buildings. Most of these buildings and features still exist and have attracted the attention of English Heritage, who have noted many of them as items of interest, just short of a full listing. The history of the site is clear as soon as you approach the gates at the top of Lippitts Hill, across from the pub called the Owl – a popular watering hole for many decades of Lippitts Hill occupants.

Enemy aircraft used the Lee Valley reservoirs to point them towards the capital and the Lippitts Hill site, and the ack-ack gunners had a perfect view of them. Eventually the site was taken over by the British Pioneer Corps and turned into a prisoner-of-war camp for captured German soldiers. Many of the features, such as the gun emplacements, were buried in spoil for security reasons. The German prisoners also left their mark at Lippitts Hill: as soon as you drive in there is a carved concrete statue of a man, which was completed by Rudi Webber (Prisoner 540177) in October 1946. After the war the Americans erected a simple memorial to mark the period of their occupation. It is still situated at the south-west corner of the camp, behind some trees.

The Metropolitan Police use of Lippitts Hill has varied over the years. The site was initially used primarily for training and for many years was the heart of firearms training in particular, with multiple ranges and training props present. It is also used for the resting of police

RIGHT The memorial to US personnel dedicated and unveiled by Mrs Lewis Douglas, the wife of the US Ambassador, is hidden in a corner of the Lippitts Hill site.

horses, a practice that continues in one of the paddocks that sits adjacent to the base. It is amazing how after a matter of hours the police horses, effectively on their holidays, simply ignore the helicopters taking off and landing nearby. Dog and other specialist training has also taken place over the years, but today the site is largely unoccupied: NPAS London is the primary occupant in the former Met ASU accommodation and hangars.

The site chosen for the ASU permanent home was developed from the site that was used for the Metropolitan Police trials with AAC Sioux helicopters. A new hangar was built at the north end of the ASU site: this, the north hangar, had sufficient room for two Bell 222s and also had workshops, offices and a control room on the first floor overlooking the pad. This single multi-purpose hangar provided insufficient accommodation, and use was made of the four nearest camp buildings for classroom, canteen, engineering and storage. Some years later a second and much larger hangar was built to the south of the apron, and the apron was extended. The south hangar has a large crane and significantly more storage and working space for the engineering team.

ABOVE Lippitts Hill crews in 1984. *(Brynn Elliott, Air Support Archive)*

LEFT Bell 222 G-METB flying over the Canary Wharf development when it was under construction. *(Brynn Elliott, Air Support Archive)*

BELOW G-SEPA sporting its new paint scheme is pictured next to G-SEPB at Lippitts Hill after its return to service. Both helicopters are fitted with an SX16 Nightsun and FLIR Leo 400 sensor turret.

The accommodation has never lost its wartime military feel, and crews have over the years passed on stories of ghostly sightings.

The site has housed three fleets of helicopters, starting with the three Bell 222 helicopters that took off from the pad and headed for the gap in the trees. Later the AS355N Twin Squirrels took off and landed using a helipad profile that was required for safety, regardless of the site's size. The lack of a clear area or runway to maximise take-off weight has always been seen as a limiting factor of the site. In the mid-2000s the arrival of the EC145 meant that the field to the south of the site had to be converted into a clear area runway, the space needed being a strip of level ground 300m long and at least 30m wide. This was so that the new helicopter could take off at maximum all up weight with a fuel load that allowed over two-and-a-half hours of flying time. A civil engineering company was contracted to build the runway and after several months Lippitts Hill became fully functional with a clear area departure – safer and more operationally effective than the helipad.

Lippitts Hill has been a much-loved home for generations of Met pilots, engineers and

ABOVE Engineers work on the rotor head of the Bell 222 at Lippitts Hill during the 1980s.

BELOW One of the Met's Bell 222 helicopters is maintained by engineers from the Met's transport department in the North hangar at Lippitts Hill. At the time the hangar had space for two Bell 222 helicopters, each with two rotor blades, so it was possible to stow them side by side.

BELOW The rear workstation of the Bell 222 helicopter, showing display and video recorder. The Bell 222 had no police role equipment in the front of the cockpit, with everything operated from the rear cabin. This is in stark contrast to the modern helicopter role fits.

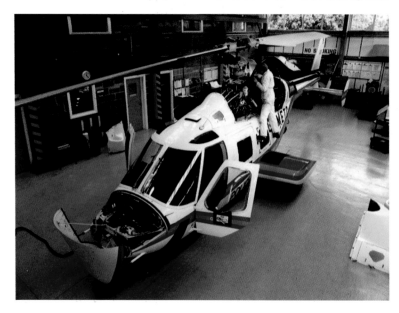

ABOVE The Met's three AS355N Twin Squirrel helicopters at Lippitts Hill.

TFOs, who all share fond memories of life at the base amid Epping Forest. When the air support relationship with Lippitts Hill comes to an end in late 2019, the future of the site will be in the hands of developers and English Heritage. It is dearly hoped that much of the history and character of the site will be retained, whatever the future holds.

RIGHT The entrance to Lippitts Hill Camp, with the two wartime defence bolsters clearly visible; these have been identified as being of historical interest.

Chapter Three

Anatomy of a police helicopter

The procurement of a powerful and capable twin-engine helicopter is just the start when it comes to building a police helicopter. It is the integration of police role equipment such as daylight and infrared cameras, tactical mapping, video displays, video recorders, police radios, digital video downlink and a searchlight that turns it into a valuable crime-fighting resource.

OPPOSITE The Eurocopter EC135 G-ESEX belonging to Essex Police pre-NPAS. It is seen in the high skidded variant with the police role equipment mounted via a demountable pod nicknamed the 'MacPod'. This enabled all the role equipment and supporting electronics to be mounted externally, making installation easier. The camera sensor (typically a FLIR) was mounted at the front and the searchlight at the back. The MacPod invariably did its job well but was unpopular owing to its weight. This helicopter is now G-POLF in service with NPAS at Lippitts Hill.

RIGHT The EC145 G-MPSA in service with the Metropolitan Police in London, flying over the Thames. The three MPS EC145 helicopters are still in service today with NPAS London.

BELOW The EC145 helicopter has a cabin that allows a range of different seating configurations. Here it is configured to carry up to four passengers in addition to the crew, using three utility seats and a club seat. This configuration is frequently used as it provides the best compromise between passenger comfort and operational versatility.

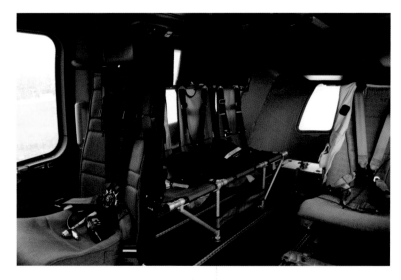

UK police helicopters

By far the most popular police helicopter type in the UK is the EC135 helicopter manufactured by Airbus (previously Eurocopter). This is a light twin-engine helicopter with a modern avionics suite, a flight endurance of over two hours with full mission equipment and the ability to carry a crew of three plus a passenger. The EC135s are joined in the NPAS fleet by the larger and more powerful EC145 helicopter, which is also known as a BK117-C2. This is in service at NPAS London and NPAS Exeter and also with the Police Service of Northern Ireland. It provides an increased capacity and flight endurance over the EC135,

and brings an extra dimension to NPAS operational capabilities.

Anatomy of a police helicopter

The EC145 is a light/medium twin-engine utility helicopter that was developed jointly by Kawasaki Heavy Industries of Japan and Eurocopter to update the already popular and capable MBB/Kawasaki BK117-C1. It was branded in Europe as the EC145, but in Japan remained as the BK117-C2. The upgrade effectively added the modern glass cockpit and avionics of the Eurocopter EC135 and removed some of the door and cabin pillars to create a large open cabin space of 9m³. This update, along with a number of other enhancements, produced a powerful, stable and capable helicopter that has the ability to carry up to nine passengers plus two crew in different seating configurations. The EC145 is in service today with a number of police, medical and security services around the world, including the French Gendarmerie and Sécurité Civile, and Swiss Air Rescue in Rega. It is also used in numbers by the US Homeland Security Service as the rebranded Lakota UH45.

The EC145 is 13.03m in length, has a height of 3.45m and a rotor disc span of 11m; the maximum take-off weight is 3,585kg. It is a very capable police helicopter. When

the Metropolitan Police EC145 fleet left the Eurocopter Deutschland factory in Donauwörth it weighed in at 1,792kg and offered a similar working payload to reach its maximum take-off weight. The process of converting the green airframe into a role-equipped police helicopter added significant weight as the mission processors, displays, workstations, radios and the camera system were integrated. This process added around 100kg of additional wiring alone. Today the NPAS EC145 fleet has a maximum variable payload of around 1,000kg, which provides for a typical mission endurance with a standard crew of pilot and two TFOs of two-and-a-half hours plus. With a cruising speed of around 150mph, it is capable of covering large distances quickly. In the majority of police tasks the variable payload after the crew is accommodated is used for fuel uplift, meaning that with maximum fuel load the EC145 can still carry a couple of passengers at 100kg each with no loss of flight endurance.

The EC145 police role fit was designed to maximise the benefits of this utility helicopter, and all the police role equipment can be removed to enable the helicopter to be re-purposed for a variety of missions. The requirement was that this should be simple, with the helicopter able to be changed from a general policing role to a specialist utility role fit in 15 minutes. This was tested at acceptance by a team of engineers who worked against

the clock like a Formula 1 pit crew. The EC145 has a high tail boom, which means crews can take advantage of rear clamshell doors that open up opportunities for easily loading and unloading equipment. With the police role equipment removed, the rear cabin of the EC145 can accommodate a significant quantity of equipment, and trials have shown this allows the transportation of urban search and rescue

ABOVE At the start of each shift and after each flight the helicopter is refuelled and made ready for immediate deployment when the next task comes in. Here Met EC145, G-MPSB, is seen on the helipad ready to fly.

BELOW The rear clamshell doors are open on the EC145, revealing the rear equipment rack which weighs in at around 75kg and can be removed for specialist transport roles. The standard role fit has three Martin-Baker passenger seats fitted across the cabin behind the tactical commander's seat.

RIGHT The rear cabin of the EC145 seen through the rear clamshell doors, with the helicopter in hybrid role fit. The utility seats and rear-facing crew seat are visible. The rear workstation, tactical commander's seat and rear equipment rack can be easily and quickly removed if required.

kit, explosives detection apparatus and other scientific equipment.

The cabin floor rails are used to accommodate a variety of seating configurations, with the general policing role fit using the floor rails to secure the tactical commander's workstation and a Martin-Baker crashworthy crew seat. This installation allows the tactical commander safe and secure freedom of movement in the rear cabin, as the crew seat can be repositioned or rotated in flight to provide comfort and operational functionality. The standard role fit has three additional Martin-Baker passenger seats fitted across the cabin behind the tactical

commander for ad hoc passenger carriage. The EC145 is a versatile helicopter and up to eight of these Martin-Baker passenger seats can be installed if required, to convert the helicopter to a more VIP passenger transport fit. The carriage of specialist resources such as CTSFOs, however, requires the use of less comfortable utility seats: these are installed in two banks of three seats attached to the floor and side mounting rails of the rear cabin. The sliding rear doors on the port and starboard sides of the cabin create an opening through which CTSFOs can be deployed quickly and efficiently upon landing, hovering or by fast rope. The three Met EC145s came fitted with hard mounting points on the exterior of the fuselage that can accommodate a hoist or an ECMS Aviation Systems fast rope and abseil beam. The EC145 can be flown in a number of hybrid roles that allow rapid redeployment should the need arise.

Powerplant

The EC145 is powered by two Turbomeca Safrane Arriel 1E2 turboshaft engines which deliver shaft horsepower (shp) output rather than just thrust. Each of the two engines delivers a take-off power of 738shp and a cruising power of 692shp. The Arriel engine is popular in helicopters because of its ability to deliver sustained power and reliability, as well as its comparatively small size and light weight. The Turbomeca Arriel family of engines have powered a variety of different helicopter models and types, and between them have amassed some

BELOW A drawing that explains the modules/stages of the Turbomeca Arriel 1E2 engine.

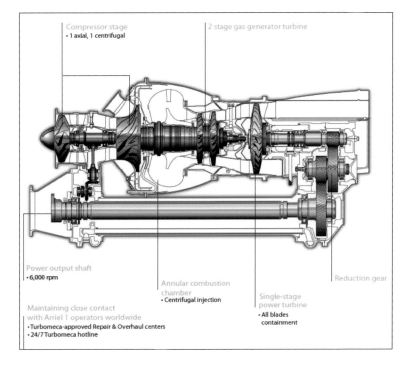

Compressor stage
• 1 axial, 1 centrifugal

2 stage gas generator turbine

Power output shaft
• 6,000 rpm

Annular combustion chamber
• Centrifugal injection

Single-stage power turbine
• All blades containment

Reduction gear

Maintaining close contact with Arriel 1 operators worldwide
• Turbomeca-approved Repair & Overhaul centers
• 24/7 Turbomeca hotline

50 million flight hours since their introduction. As part of the Arriel 1 family, the 1E2 features a design centred around five modules for ease of maintenance. The engine comprises a single-stage axial compressor, a single-stage centrifugal compressor, a two-stage gas generator turbine and a single-stage power turbine. The five modules combine to deliver a powerful, safe and reliable engine that can fly for many hours between major maintenance activities and is the perfect powerplant for the high hours requirements of police helicopters. NPAS has a power by the hour (PBH) contract to assist in managing engine maintenance and guaranteeing availability of spares. PBH is effectively an insurance policy whereby the operator pays for every hour flown at a fixed rate. The engines used on the Met's EC145 helicopters were each flying around 1,000 hours pre-NPAS and the PBH contract between Airbus and engine manufacturer Turbomeca ensured that if any one of the five engine modules needed maintenance a complete new engine was delivered ready for the Met's engineers to install at Lippitts Hill; this continues today with NPAS at Airbus in Oxford.

Rotor system

The update of the BK117-C1 into the C2 variant (EC145) saw the adoption of carbon-fibre components in some areas to reduce weight and

noise and to increase performance. One of these areas was in the manufacture of the four main rotor blades, a technological development taken from the EC135. The EC145 has a fully rigid rotor system, in contrast to the semi-rigid rotor head fitted to the Met's previous AS355N Twin Squirrel helicopters. This produces a firm and responsive ride, meaning the helicopter can turn very quickly and respond almost instantly to the pilot's control inputs. This caused some concern for crews that transferred from another helicopter type, as some believed the ride would be too hard and vibration would be an issue. There is no doubt that the AS355 was exceptionally comfortable to fly, but

ABOVE The Arriel 1E2 engine.

LEFT One of the two Arriel 1E2 engines fitted to the EC145 helicopter, with the engine cowlings removed.

ABOVE The main rotor head is machined from a solid piece of titanium that is attached to the rotor mast with 12 titanium studs and special nuts. The pitch links use commands from the collective lever to change the pitch of the rotating blades to provide more or less lift as power increases or decreases. The cyclic control changes the position of the entire rotor head assembly to move the rotator disc and move the helicopter forwards, backwards, left or right.

the fears were unfounded and the EC145, like its EC135 sister type, is exciting, dynamic and fun to operate.

The rotor head is machined from solid titanium, and the four blades are attached and controlled via a series of titanium bolts and rods and a number of pitch links, which control the pitch of each blade as the pilot applies adjustment for

power and lift and to the cyclic control for pitch and roll. The power delivered to the main rotor by the Turbomeca Arriel 1E2 engines and the Kawasaki-manufactured main rotor gearbox is matched using a variable rotor torque matching system (VARTOMS). In simple terms, VARTOMS aims to match the torque delivered by both engines and vary the rotor speed based upon a number of factors, such as airspeed, altitude and outside air temperature. It does this by setting the torque of the right engine with reference to the left engine and aims to provide maximum efficiency at all times. This system has been credited with reducing noise so that the EC145 is one of the quietest helicopters in its class.

Fuel system

The EC145 uses Jet-A1 fuel to power its two engines. The fuel storage system, comprising two connected main fuel tanks and two engine supply tanks, is fitted beneath the cabin floor. Fuel uplift is typically calculated in kilograms of Jet-A1 fuel rather than litres, with the capacity of the main and supply tanks being a total of 867.5 litres. The density of Jet-A1 fuel varies according to the ambient temperature: typically at 15°C a litre of Jet-A1 weights 0.804kg. The two Arriel 1E2 engines typically burn fuel at a rate of 2kg per minute each, meaning that a calculation of 4kg of fuel per minute of flight is used for general planning purposes.

The fuel is transferred from the main fuel tanks to the two supply tanks, which are forward of the main tank and then feed to the two engines. For engine and rotor start, booster pumps are used to deliver fuel to the engine from the supply tanks. Once the engine is up and running, however, the booster pumps are switched off and the engines effectively suck through the required fuel. The

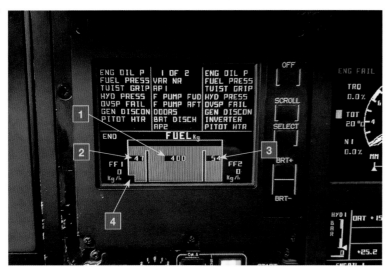

LEFT The fuel state is displayed on the EC135 and EC145 as part of the Caution Advisory Display.

1 Main fuel supply showing 400kg

2 Fuel supply tank No 1 engine (47kg)

3 Fuel supply tank No 2 engine (54kg)

4 Note the visual indication that No 1 supply tank has slightly less fuel (8kg or 2 mins flying time) than No 2 supply tank.

LEFT Master switches for the avionics suite and fuel system are in the roof console switch panel, which is located above and behind the pilot and front TFO seats in the EC145.

1 Avionics master switches
2 Fuel transfer pumps fore (F) and aft (A) used to transfer fuel from main tanks to supply tanks
3 Fuel primer pumps 1 and 2 used for engine start.

transfer of fuel from the main to the supply tanks is achieved by fuel transfer pumps, which are switched on and off by the pilot, using switches in the roof switch console. The supply tanks hold 101kg of fuel between them, with the No 1 engine supply tank holding 47kg and that for No 2 engine holding 54kg. This 8kg difference at a typical burn rate of 4kg per minute equates to two minutes of extra fuel. In the highly unlikely event that the helicopter runs exceptionally low on fuel, the No 1 engine flames out two minutes before the No 2, a safety feature of last resort. The fuel transfer pumps continue to run when the main tanks are empty unless they are switched off; sensors indicate that they are running dry and the pilot is notified via a message that appears on the caution advisory display (CAD) in the centre of the instrument panel, alerting the pilot to switch off the fuel transfer pumps to avoid pump damage.

The helicopter is fuelled via an external refuel point, and a highly accurate digital fuel gauge is displayed graphically as part of the CAD. A police helicopter has a minimum permitted landing allowance (MLA) of sufficient fuel to fly for 20 minutes by day or 30 minutes by night; that is, 80kg or 120kg of fuel respectively in the EC145. During flight the crew calculates the helicopter's endurance by adding together the content of the main and supply tanks in kilograms, dividing the total by four and then taking off the MLA to deliver an endurance in minutes. This is always verified by the pilot and is monitored throughout the flight, as the rate of fuel burn can vary

depending upon the type of flying undertaken and factors such as wind speed.

In the event of an engine fire it is possible to instantly shut off the fuel supply to either engine using a gated emergency fuel cut-off switch.

BELOW The top section of the instrument panel of the EC145, featuring the fuel emergency cut-off switches, engine fire extinguishers, warning unit and additional instrumentation, such as clock, airspeed indicator, standby attitude indicator, barometric altimeter and engine speed indicators.

1 Engine fire extinguisher button. Only active if fire detected. First press puts bottle 1 into No 1 engine, second press puts bottle 2 into No 1 engine
2 Emergency fuel cut-off switch for No 1 engine. Lift gate and press to immediately close a fuel supply line valve cutting off fuel
3 Switch replicated for No 2 engine.

ABOVE The bottom section of the EC145 instrument panel, pictured in flight. This shows the battery master switch that is used to apply master power and controls for the No 1 and No 2 generators. The engine start switches are above that and the CAT-A switch (High N^R), which is used to increase rotor speed to maintain performance in certain phases of flight.

BELOW At the rear of the centre console in the EC145 are the switches that apply power to the VMS and the camera system (MX15). These were moved backwards from the slant panel to enable a swap with the ELT activation switches after a couple of inadvertent activations.

The fuel cut-off switches for the No 1 and No 2 engines are located to the left and right of the warning unit at the top centre of the instrument panel. Its protective red gates are secured with a wire telltale, and when they are lifted the securing wire is broken, which allows the switch to be pressed; this immediately closes the fuel cut-off valve for that engine. Adjacent to the cut-off switches are engine fire extinguisher switches, which allow the pilot or crew to fire either one or both of the two engine bay fire extinguishers in the event of a fire.

Electrical system

The EC145 electrical system operates at 28v direct current (DC) and electrical master power is provided via a master battery; this is located high on the starboard side of the helicopter adjacent to the engine and gearbox, and is accessed for maintenance via a small hatch. The battery master switch in the centre of the instrument panel provides power sufficient for the pilot to commence engine and rotor start, using the engine start switches located to the left and right of the battery master switch. Once the engines are powered up the two onboard generators deliver power at a clean and stable 28v DC. This stability is required so that all the sensitive avionics and police mission systems operate effectively. The master battery powers the helicopter's electrical systems on the ground, but it soon runs out of power – and as such an external power supply unit, referred to as ground power, is used for any ground-based testing, training or activities that require powered avionics and role equipment. In flight the loss of engine power causes the generators to fail, and the master battery supplies power to critical flight systems so the pilot can safely complete an autorotational landing.

Once the generators are up to power and stable, the pilot applies power to the avionics suite via a master switch located in the centre roof console. At the same time the pilot instructs the front seat TFO that power can be applied to the police mission equipment, using the video management system (VMS) and MX15 switches located at the rear of the centre console. These switches enable the police mission equipment to be isolated and to ensure it is only powered on once the generators are running and delivering the required voltage. Failure to do this causes

RIGHT The collective control lever of the EC145 with the throttle twist grips (with different grip patterns) that control No 1 and No 2 engines, and the circuit breakers visible below.

1 Throttle twist grip No 1 engine
2 Throttle twist grip No 2 engine.

the equipment to fail to power up correctly and therefore fail to operate correctly during the flight. A series of pop-out fuses located in the centre roof console and beneath the collective lever enable the pilot to troubleshoot in-flight and address equipment or avionics failures.

Flight control and avionics

One of the primary developments of the EC145 as it grew from the BK117-C1 was the addition of a full glass cockpit, all flight instruments now being digital and accessible via a range of displays. The glass cockpit has two Thales SMD45 primary flight displays, which are controlled via control switches in the centre console and deliver full redundancy if either display fails. It is possible for the pilot to switch the content of each display and composite critical flight information on to a single unit. A video radar unit enables other information to be interfaced on to the flight displays, and in the EC145 this comes from a weather radar that is

LEFT The cockpit of the EC145 is built around a glass cockpit that comprises two digital primary flight displays, a digital CAD and vehicle and engine management displays (VEMD). It is pictured here in the SPIFR police role fit, with a police mission display in the front left position.

RIGHT The EC145 is a full dual-pilot IFR-certified helicopter, and police mission equipment can be removed to reinstate a portrait Thales SMD68 primary flight display, which replicates the landscape Thales SMD45 primary flight displays fitted in the front right position. Here the dual-flight controls are fitted in dual-pilot mode.

RIGHT The top of the collective pitch lever, showing the various switches and controls available to the pilot.

1 Collective lever switching to control landing lights power, steerable landing light and, if required, the SX16 searchlight.

fitted as an option. This is responsible for the rather bulbous nose cone, and proves extremely useful when operating on instruments: it gives the pilot a clear view of the weather conditions and specifically any nearby storms.

A three-axis autopilot that can control pitch, yaw and roll can be used to assist with maintaining heading, height and speed. The recently uprated EC145 T2 has a four-axis autopilot: the fourth channel can access the collective lever to control power. All police helicopters are required to have an autopilot owing to the high pilot workload and the dynamic nature of the operational role and environment. The collective pitch lever controls the power transferred to the rotor blades and varies their pitch to change lift. It contains the two throttle twist grips, each with a different grip pattern to enable the pilot to easily differentiate between them. The throttles are manually twisted from off through to start, idle and then on to flight positions, and are set as required for the flight with fuel flow managed electronically. In the event of a fuel management system failure the throttles can be operated manually to control fuel flow to each of the engines. A safety system prevents the accidental moving of the throttles to the off position, and the pilot must take a positive action to press a safety button that moves the throttle beyond the safety stops. The top of the collective pitch lever contains a number of controls and switches that allow the pilot to control the steerable landing light and police searchlight if required or to operate the windscreen wipers, among other things. The cyclic pitch control is used to control the direction of travel by varying the position and angle of the rotor head, moving it forwards, backwards and sideways. It also contains a variety of trim and other switches and controls, which enable the pilot to perform key functions in flight. A radio press to talk (PTT) button is also provided, which enables the pilot to talk on any selected radio. The final flight controls are the tail rotor control pedals, which control the rotation of the helicopter left or right by increasing or reducing the power of the tail rotor; this is countering the rotational effect of the main rotor head to prevent the cabin counter-rotating.

Police helicopters fly under public transport safety standards for most missions, meaning that the same level of safety is afforded to the crew in the event of engine failure as applies to the public flying on a commercial airline. In the event of a sudden loss of power a single Arriel 1E2 engine can provide sufficient power for the helicopter to continue flying and land safely, without injuring the passengers and crew or damaging the airframe in any way. To ensure additional power is available for take-off and landing a CAT A mode can be selected, which raises the rotor speed (N^R) slightly to provide extra power to maintain CAT A performance. On the EC135 this is sometimes referred to as a High N^R setting. The CAT A switch is located on the instrument panel below the CAD, and is selected for take-off and landing phases of flight or as required by the pilot.

Instrument panel

The EC145 is certified as a single-pilot instrument flight rules (SPIFR) helicopter. Redundancy is built in, ensuring that the flight instruments remain operative at all times,

BELOW The cockpit of the EC145 in police role fit.

1 Primary flight displays – Thales SMD45
2 Traffic Collision Avoidance System (TCAS)
3 Extra Analogue Radalt – UK police requirement
4 Caution Advisory Display (CAD)
5 VEMD displays
6 Garmin GNS 430 ATC radios and nav aides
7 Transponder
8 Primary flight display controller
9 Automatic Direction Finder – Nav Aide
10 Video radar unit controller
11 NAT A320 station box – radios and ICS
12 Chelton CH450 Police Radio Controller
13 Slant Panel Annunciator and role equipment power switches
14 Master battery switch
15 Engine start switches 1 and 2
16 CAT A switch
17 No 1 engine emergency fuel cut-off and fire extinguisher switch (gated for safety)
18 Warning unit
19 Clock, airspeed indicator, standby attitude indicator (AI), barometer, altimeter and triple tachometer
20 Front TFO mission video display
21 Front TFO PTT
22 Wescam MX15 hand controller
23 Collective lever
24 Yaw pedals.

ABOVE The EC145 cockpit instruments pictured at night, in flight over London.

something critical for IFR flying. The dual controls and co-pilot's primary flight display, a Thales SMD68 portrait display, are removed when the helicopter is in standard police role fit but replaced for pilot training and examination. The instrument panel comprises a range of digital displays that replace the variety of round analogue instruments found in many older helicopters and are present in the Met's AS355N fleet. This digital cockpit is referred to as a glass cockpit. In the EC145 the two primary flight displays are complemented by a suite of LCD displays that comprise a central CAD, which is used to display fuel state and cautionary warnings that require action. The CAD is joined by a pair of vehicle and engine management displays (VEMD), which display critical information to the pilot. The digital displays use a first limit indicator (FLI) system,

LEFT The centre of the EC145 instrument panel, showing the warning unit, CAD and VEMD.

1 VEMD display showing first limit indicator (FLI)
2 VEMD display showing engine oil temperatures and pressures along with hydraulics pressures and transmission oil status
3 Caution Advisory Display and fuel state gauge
4 Master battery switch
5 Generator switches
6 Engine start switches
7 CAT A – High N^R switches.

RIGHT The cockpit instrument slant panel of the EC145, showing pilot and police radios and audio and display controllers.

1 Garmin GNS 430 pilot radios and navigational aides
2 Garmin transponder
3 Primary flight display controls
4 ELT emergency activation switch
5 Downlink actuator annunciator
6 Chelton CH450 radio control head
7 NAT station box for front TFO controlling radios and intercom (ICS).

which considerably simplifies engine and torque monitoring by the pilot. The system monitors various systems and displays the first of these that will reach an operating limit. The FLI uses a clear colour display to indicate this, with limits clearly identified. This innovative flight instrumentation means the pilot is always aware of when the first limit will be reached and can manage the flight accordingly. The VEMD is also used to display engine, oil and hydraulics temperatures and pressures, along with generator status and outside air temperature, in a simple easy-to-read form.

To complement the digital displays there is a variety of other instrumentation. This includes an analogue radar altimeter, airspeed indicator, standby attitude indicator (referred to as an artificial horizon), a clock and a triple tachometer. A traffic collision avoidance system (TCAS) has also been integrated, and

LEFT The instrument panel, showing the pilot's primary flight instruments.

1 Thales SMD45 primary flight displays 1 and 2
2 Traffic collision avoidance system (TCAS)
3 Flight checklists (pre-flight, after take-off and pre-landing checks)
4 Analogue radar altimeter
5 Vehicle and engine management displays.

1 Pilots NAT A320 station box – radios and ICS
2 Original digital downlink control panel
3 Controller for co-pilot primary flight display (Thales SMD68) when fitted
4 Primary flight display master controller used to composite or switch displays in event of display failure
5 Weather radar controller
6 Skyforce Observer mapping control head
7 Skyshout public address system controller
8 Engine vent bleed valves
9 Skyquest dual video controller
10 Skyquest evidential and replay video recorder
11 Auxiliary video input panel
12 Cockpit voice and flight data recorder controller
13 SX16 Nightsun controller
14 Cockpit fire extinguisher No 1
15 Collective lever. Note different grip patterns on throttle for No 1 and No 2 engines.

this displays information about any air traffic that is presently transponding in the airspace around the helicopter. The range can be varied according to operational need, and the TCAS display gives a bearing and height indication to assist the crew in making a visual identification.

Communications

A constant challenge is how to manage the audio communications required by the crew. The audio has to be multiplexed from up to seven different sources, allowing each crew member to control the volume of an individual radio. The pilot, for example, has a priority focus upon the ATC radios and is less interested in the police radios, but they still need to communicate with the crew via the intercom and gain situational awareness from this source. Audio supplied to the Alpha aircrew helmets is controlled via an audio mixer unit known locally as a station box. This contains a number of buttons and switches that resemble a recording studio mixing desk, and these allow individual selection of radio and intercom volume and PTT options. It is possible therefore to listen to the six radios, talk on the intercom and transmit on any one of the radios using this single box.

LEFT The tactical commander's rear workstation on the EC145.

1 Tactical commander's NAT A320 station box – radio and ICS
2 Chelton RH450 police radio controller
3 Tactical commander's primary 15in mission video display
4 Tactical commander's secondary 15in mission video display
5 Mapping keyboard
6 Tactical commander's PTT switch
7 Martin-Baker crew seat
8 Connectors for rear workstation; shunt plugs stowed on rear workstation are used when it is removed
9 SX16 Nightsun controller.

BELOW The cyclic control lever provides the pilot of the EC145 with access to radio PTT and a variety of other functions.

1 Reset switch
2 Re-engage 'SAS ON' switch
3 SAS or Autopilot Cut
4 Beep trim
5 FTR.

LEFT The front of the centre console of the EC145 after delivery in 2007. This has been changed over the last decade to accommodate mission equipment upgrades.

ABOVE The EC145 port side.

1. Weather radar
2. Spectrolabs SX16 Nightsun
3. Downlink transmit antenna and actuator
4. Sliding rear door – can be jettisoned if required via red handle inside or outside cabin
5. Front door – can be jettisoned if required. Orange door markings denote passenger doors
6. Rear clamshell doors open to give access to rear cabin
7. Radio antennas
8. Refuelling hatch.

BELOW The EC145 starboard side.

1. Wescam MX15 camera sensor
2. NAT Skyshout public address system
3. Gigawave uplink video receive antenna
4. Front door – can be jettisoned if required
5. Rear sliding door
6. Rear clamshell doors
7. Radio altimeter antenna – RadAlt
8. Master battery hatch
9. External (ground) power socket hatch.

It allows the crew to turn audio on and off for each radio input and to control individual radio volumes, as each individual pilot or TFO has their own preference regarding the prominence of the audio presented in their helmet. Aircrew helmets are the most expensive item of equipment that the pilot and TFOs are issued with, as they are custom fitted and equipped with powered active noise reduction (ANR). The audio and intercom system also allows the pilot to isolate the rest of the crew should police communications become dominant and impede communications via ATC radios; this is achieved via an audio control panel in the centre console.

Station boxes

The individual audio for the pilot and TFOs is managed using an NAT A380 station box; the pilot and front seat TFO boxes are located within the centre console and the one for the tactical commander is located above the work station adjacent to the Chelton Radio control head in the rear cabin. It is only possible to transmit on one radio at a time. The PTT selector is a push button that backfills green to show the selected radio (ATC1, ATC2 or police tactical radios TAC1-5). Each crew member needs to be aware of what they have selected and ensure that their transmission is on the correct radio.

The final piece of the audio system is the intercom function of the crew helmet microphones. The volume of this is simply selected by turning up or down the rotary switch, and as with other tactical radios it is possible to mute by pressing the intercom button, which turns it off. The quantity and clarity of cockpit communication required for effective police operations means that the preference is usually for a live or 'hot' microphone, meaning it is always open and conversation via the intercom is free and easy. The placement of the microphone very close to the lips reduces background noise significantly, and this means communication is clear and effective.

Station boxes are also provided for the rear seat passengers, who typically wear headsets rather than helmets. A range of audio interface leads means that the crew can control their ability to transmit on radios or speak on the intercom, which cuts down unnecessary voice traffic. The audio system is configured to

ABOVE The EC145 has external mounting points, which were factory installed to allow the fitting of an external hoist or rope/abseil beams.

BELOW The NAT-manufactured station box enables crew members to each take control of the audio distributed into their aircrew helmet or headset.

1 Radio Transmit Selector. Push to select, selected TAC shows green
2 Intercom (ICS) volume control
3 ICS microphone control VOX (voice operated) or HOT (live)
4 Individual TAC radio controls. Push to turn ON and rotate for volume control.

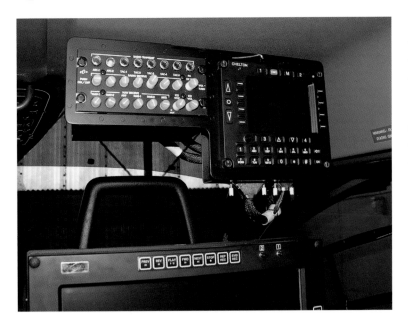

ABOVE The tactical commander's audio suite includes a station box and a Chelton RH450 radio control head. Using this, the tactical commander can control the four police radios and select which one to talk on using the PTT selector on the station box. Most of the in-flight police communication with the ground is done by the tactical commander.

PTT switches

The EC145 has fixed-position PTT switches for the TFOs: these have a positive click for PTT that can be felt through flying boots. The PTT switch is in a fixed position after concerns were raised about using the more common pedal-type switches; these use a long fly lead to allow the TFO to move their position around the cabin floor. Experience shows that when things get hectic in the cockpit and the crew are working hard, the last thing they need is a wayward radio PTT switch. The pilot has a PTT switch built into the cyclic pitch control, and an alternative is provided via an audio isolation panel in the centre console. This ensures redundancy and guarantees communication ability.

Air traffic radios

The EC145 is fitted with two Garmin GNS 430 GPS navigation and communications boxes, which allow the pilot to communicate with ATC on the required frequency using ATC1 and ATC2 Garmin boxes. The GNS 430s are fitted in the slant panel and are referred to as 'box 1' and 'box 2' when communicating within the cockpit. The pilot typically retains communications with the ASU base using the company radio frequency on one radio box with the appropriate ATC frequency on the other. The GNS 430 dataset is routinely updated. The presence in the helicopter of police tactical mapping assists greatly with navigation, but the dataset used is not cleared for flight planning and management at air traffic and airspace level, so it can only ever be used as a cross reference by the pilot. In recently upgraded EC135 T2 helicopters NPAS have exchanged the ATC radios to the more capable Garmin GTN 650 radio navigational system.

BELOW The tactical commander's radio PTT is built into a fixed position and uses a positive click switch that can be felt through flight safety boots.

provide intercom and radio communications via a separate amplifier located in the rear avionics shelf, and this is controlled using an NAT AA35 control panel mounted in the roof of the rear cabin. This allows the tactical commander either to isolate the rear passengers, allowing them to talk freely without interfering with cockpit communications, or to bring them into conference with the rest of the crew.

Police radios

The EC145 is fitted with four Thales vector radios that operate on the emergency services Airwave network. These radio units are mounted in the rear avionics shelf, which has been modified to accommodate four police radios along with automatic direction finding (ADF), distance measuring equipment (DME) and TCAS processors. The installation of the various radio antennas presented a challenge; accommodating two ATC VHF radio antennas,

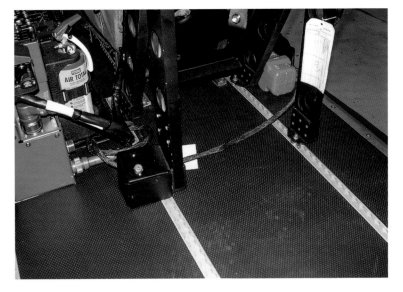

five police radio Airwave antennas along with GPS, TCAS, NDB/ADF and other antennas without interference was not easy given the limited space available. Following a significant amount of testing, the police radio antennas (perhaps the only antennas not conventionally fitted to a helicopter) found space on the tail boom and forward belly panel. Operational use of the various police radios, however, demonstrated that some appeared to be more reliable than others – and this was attributed to the location of the antenna.

ABOVE LEFT The more standard radio PTT is a pedal type connected to a fly lead. This can be moved around the cabin and is secured by Velcro.

ABOVE The NPAS EC135 T2 fleet has recently been upgraded and equipment now includes a Garmin GTN 650 radio navigational aid, pictured here on the slant panel.

The police radios are each loaded with a fleet plan that effectively creates a directory of radio talk-groups that are available to the crew. A great deal of planning goes into determining the

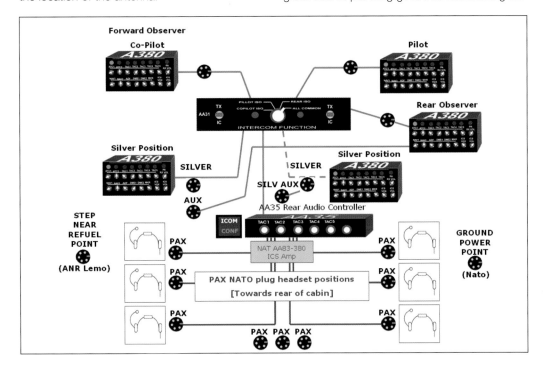

LEFT A schematic showing the components of the audio and ICS system.

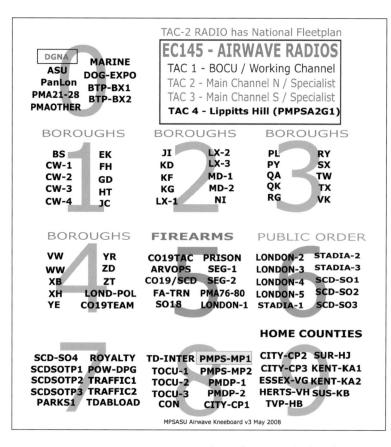

DGNA		**MARINE**
ASU		DOG-EXPO
PanLon		BTP-BX1
PMA21-28		BTP-BX2
PMAOTHER		

TAC-2 RADIO has National Fleetplan

EC145 - AIRWAVE RADIOS
TAC 1 - BOCU / Working Channel
TAC 2 - Main Channel N / Specialist
TAC 3 - Main Channel S / Specialist
TAC 4 - Lippitts Hill (PMPSA2G1)

BOROUGHS

BS	EK
CW-1	FH
CW-2	GD
CW-3	HT
CW-4	JC

BOROUGHS

JI	LX-2
KD	LX-3
KF	MD-1
KG	MD-2
LX-1	NI

BOROUGHS

PL	RY
PY	SX
QA	TW
QK	TX
RG	VK

BOROUGHS

VW	YR
WW	ZD
XB	ZT
XH	LOND-POL
YE	CO19TEAM

FIREARMS

CO19TAC	PRISON
ARVOPS	SEG-1
CO19/SCD	SEG-2
FA-TRN	PMA76-80
SO18	LONDON-1

PUBLIC ORDER

LONDON-2	STADIA-2
LONDON-3	STADIA-3
LONDON-4	SCD-SO1
LONDON-5	SCD-SO2
STADIA-1	SCD-SO3

HOME COUNTIES

SCD-SO4	ROYALTY	TD-INTER	PMPS-MP1	CITY-CP2	SUR-HJ
SCDSOTP1	POW-DPG	TOCU-1	PMPS-MP2	CITY-CP3	KENT-KA1
SCDSOTP2	TRAFFIC1	TOCU-2	PMDP-1	ESSEX-VG	KENT-KA2
SCDSOTP3	TRAFFIC2	TOCU-3	PMDP-2	HERTS-VH	SUS-KB
PARKS1	TDABLOAD	CON	CITY-CP1	TVP-HB	

MPSASU Airwave Kneeboard v3 May 2008

ABOVE The TFO's radio kneeboard guides the TFO quickly to the required radio talk-group.

most appropriate talk-groups for installation, as the radios have a limited memory capacity. The four radios are controlled via a Chelton CH450 control head that is mounted in the helicopter slant panel to the left of the pilot's Garmin GNS 430 boxes. This installation location is common across the NPAS fleet and gives the front of the cockpit clear and unobstructed access to the police radios.

The way that the helicopter is operated

means that the vast majority of the police radio communications are undertaken by the tactical commander in the rear of the helicopter. The tactical commander therefore needs to have access to a radio control head to enable selection of talk-groups. This is achieved by means of a second Chelton radio control head, this time an RH450 that is effectively a slave of the master control head. The rear head is capable of all functions except master power on and off, and is a repeater of the front CH450. The control head screen is divided into quadrants, each representing a tactical radio (TAC1–4), and the talk-groups are organised into pages or folders with each folder containing eight talk-groups.

By following set procedures and rigid radio discipline, the crew are able to communicate and are always aware of what talk-group is in use. Navigating around the fleet plan is achieved by searching for a specific talk-group by using an alpha numeric search or going directly to a specific page. The Met crews use the page number method, and a TFO kneeboard has been developed to lay out at a master level where the talk-groups are located. If, for example, an ASU tasking is on the Borough of Brent, which has the call sign 'QK' and was using the talk-group QKDespatch1, then the TFO will look at their kneeboard and work out that the QK talk-groups are on a page number in the 30s, page 33 to be precise. It is then possible to go quickly to page 30, and scroll through to page 33 and the talk-group required with the minimum input. The TFO radio

RIGHT An explanation of how the radio kneeboard is used to navigate the radio head and find the correct radio talk-group.

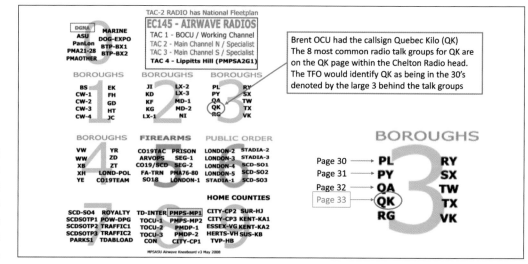

Brent OCU had the callsign Quebec Kilo (QK)
The 8 most common radio talk groups for QK are on the QK page within the Chelton Radio head.
The TFO would identify QK as being in the 30's denoted by the large 3 behind the talk groups

BOROUGHS

	PL	RY
Page 30 →	PL	RY
Page 31 →	PY	SX
Page 32 →	QA	TW
Page 33 →	QK	TX
	RG	VK

kneeboard is regularly updated to ensure that it reflects the current mix of local and national fleet plans. The move to NPAS means that the air support reach is now national, and as such rather than just having main force talk-groups for the whole country NPAS aircraft have a mix of local talk-groups for any force they are likely to support. The aim is to eventually come up with a national radio fleet plan, but this is limited at present by airborne radio capacity.

Hand-held radios

The EC145, in common with all NPAS helicopters, has a cradle-mounted radio as part of the role equipment fit. This allows for flight following by NPAS using a Sepura handset. If the helicopter lands to deal with an incident, something that happens from time to time, the officers leaving the helicopter have an ability to communicate. The hand-held radio uses an alternative antenna that shares the TAC3 position. A switch on the centre console (ALT TAC3) adjacent to the VMS and MX15 switches allows the audio system to switch to the hand-held radio as an alternative TAC3. The hand-held radio also provides resilience in the event of a radio head failure: losing the Chelton CH450 would wipe out all the police radios in one fell swoop.

Exploring the police role equipment

The helicopter camera system

At the heart of any police helicopter is the camera sensor. Police air support missions are impossible without the ability to see effectively day or night from the air. The NPAS London EC145 fleet is fitted with a Wescam MX15, and power is applied using an MX15 switch in the centre console. This powers the master control unit (MCU) and turret, and ensures that even when the helicopter is not in operational police use the stowed camera is protected. The NPAS EC135 T2 fleet fitted with the newer generation Wescam MX10 systems have all the electronics needed to operate the camera integrated within the turret itself; there is no separate MCU, which saves significant weight and means installation is less complex.

LEFT The hand-held radio is mounted into a cradle similar to ones used in a vehicle. In this photo the Motorola MTH800 hand-held radio is shown fitted; the cradle was modified to take a Sepura handset in recent years.

BELOW The Wescam MX15 hand controller.

1   The ZOOM switch is pushed up or down to zoom in or out of a target
2   The VIC switch is used to scroll through the available sensors
3   Steering controls are used to activate AUTO/MANUAL mode or engage Rate or Auto AID modes. The SLEW switch slews the camera on to a target sent by the mapping
4   The slew transducer is used to steer the turret in azimuth and elevation. Its concave top allows the TFO to rest their thumb on to the switch and rock it to steer. The harder the switch is pressed, the faster the turret movement
5   The FOCUS wheel is roated to bring the imagery into focus.

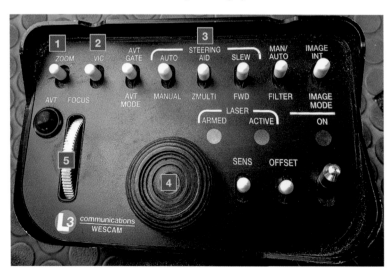

LEFT The QUAD screen display puts all three Wescam camera images together with the mapping on a single display. The TFO can press any quadrant to bring that image up to full screen; pressing the screen again returns to a quad view. In this training screen grab the mapping imagery shows the camera point of impact on the map for target identification.

CENTRE A screen capture of the EOW daylight TV camera of the Wescam MX15.

The turret is operated via a hand controller and can be steered through 360° in azimuth and between +20° and -100° in elevation, using a hand controller. The typical operating preference is to look out at around 90° to the right of the helicopter with an elevation range looking down of between -30° and -60°. The pilot aims to place the helicopter so that the turret sensors are able to look down and to the right of the helicopter at all times; this gives a good range of dynamic movement and a clear field of view. Contained within the turret are four sensors, with two of them sharing the same optics.

Electro Optic Wide (EOW)

This is a daylight TV camera with a zoom capability. It enables the crew, from a typical operating height of around 1,500ft, to view a whole housing estate or zoom into an individual house or garden.

Electro Optic Narrow (EON)

This is known as the spotter scope, and is effectively a fixed focal length telescope with a high level of magnification. It can operate in daylight as a full colour TV image or in very low light (LL) as an intensified LL image. The field of view is very narrow and maintaining situational awareness is a challenge. TFOs learn to control the EON sensor with very slight movements and to use the EOW zoom sensor to maintain situational awareness.

LEFT A screen capture of the EON spotter scope camera of the Wescam MX15.

Low Light (LL)

The LL sensor uses the same optics as the
EON sensor but replaces the colour TV camera
with an image intensifier that is capable of
producing an image in LL. The LL sensor only
needs a very small amount of ambient light from
street lighting or another source to produce a
clear image. The LL sensor creates a black and
white image; however, it does not necessarily
follow the conventions that the brain may be
used to, so dark clothing can appear light and
vice versa. One thing that the LL sensor does
deliver is the ability to obtain clothing logos,
descriptions or vehicle details such as number
plates in LL conditions.

Infrared

The IR sensor represents the level of heat
present as a shade of grey, with either black or
white being at the upper and lower end of the
greyscale spectrum. The TFO can select the
view of a scene as 'black hot' or 'white hot',
and the representation of heat shown on screen
is adjusted accordingly.

Video in Command (VIC)

The camera system delivers imagery from all the
sensors all the time, and they move together to
show the same object or scene. The sensors
each have slightly different bore sights and the
TFO learns to use the cross hairs on the wide
angle EOW imagery to queue up the detail
required before switching to the narrow EON
sensor to view it on screen. The TFO controls
the steering, focus and function of the various
sensors via a hand controller and is capable
of doing so for only one sensor at a time. The
sensor that is selected and under the control of
the TFO operating it is known as VIC.

Stabilisation and geo-referencing

The turret uses a number of gyroscopes for
stabilisation; these ensure that whatever the
movement of the helicopter, the image on
screen is clear and stable. The turret also
contains a component known as an inertial

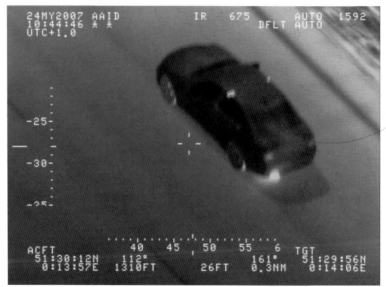

ABOVE A screen
capture of the IR
sensor of the
Wescam MX15.

measurement unit (IMU), which enables the
turret to calculate exactly where it is in space
at any moment. Through a process of data
exchange, the turret is able to calculate the
exact position that the sensor is looking at, its
height above sea level and its distance by slant
angle from the helicopter. This information is
displayed to the TFO on the screen and assists
greatly when retaining situational awareness.

The interface between the camera turret
and mapping system means that not only can
the camera calculate with a high degree of
accuracy the geographical point on earth that
the TFO is looking at, but it can also display
this on the mapping system as a camera point
of impact. This feature is of great benefit when

ABOVE A QUAD image screen grab showing the three Wescam MX15 sensors alongside the moving map display.

position of the sensor even when you take your hand off the controller. This means that if the target is a specific house or road junction, the camera will hold that position regardless of what the helicopter does and where it flies. In manual mode this would not be possible, and the sensor would drift away from the target as soon as the helicopter moved. Auto mode means that it is possible to select a specific location by map reference or by using a cursor on the map, and to send that to the camera system. It is then possible for the TFO to slew the camera directly on to that target.

managing a vehicle pursuit or a search, as the exact location of a vehicle or object can be quickly and accurately identified.

Auto or manual mode

The introduction of GPS-based technologies and the presence of the IMU mean that it is now possible to operate the camera in auto mode, which uses geo-pointing to hold the

Controlling the camera system – the hand controller

The camera is steered in azimuth and elevation via a slew transducer, with the TFO controlling the fine movement of the turret around the target using their thumb. The slew transducer on the MX15 is a round concave switch rather than a joystick, and the TFO rests their thumb on the top of it. The switch measures the amount of pressure applied as well as the direction in which it is applied: the harder the slew transducer is pushed the faster the turret responds. When following vehicles it is possible

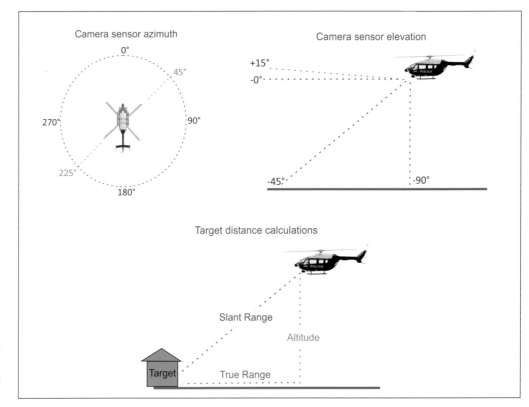

RIGHT Diagrams explaining azimuth, elevation and slant angle.

to enhance things further by using a AID mode, which replicates a specific slew transducer input until it is changed. This function, known as RAID in manual mode and AAID in auto mode, means following a vehicle travelling along a stretch of road at speed is much easier.

Zoom and focus

The other two major functions are the zooming of the sensor, a function available on the EOW TV camera and the IR sensor, and focusing the image on screen. These are achieved by using a spring-loaded switch to zoom in and out and scrolling a wheel on the hand controller for focus. As with the slew transducer, it is possible to customise the gain of these switches to slow down or speed up the rate of zoom.

ABOVE LEFT The Wescam MX10 is fitted to the recently upgraded NPAS T2 fleet of EC135 helicopters.

ABOVE The Wescam MX15.

LEFT A screen grab showing the Wescam screen symbology.

1 Date and time from GPS
2 Sensor ID. If reverse video then sensor is VIC
3 Camera elevation
4 Cross hairs
5 Aircraft position
6 Camera azimuth
7 Target position.

TFO mapping

One of the most fundamental changes in police aviation has been the introduction of accurate moving map technology, a flying satnav. In the days before moving maps pilots and TFOs operated exclusively from a mix of Ordnance Survey (OS) mapping and A–Z street map books. The availability of aviation spec mapping systems was revolutionary because it allowed crews to navigate with great accuracy around their operational areas and to have at their fingertips mapping data from building outline up to 1:500,000 aviation charts. The technology has advanced rapidly over the last decade, and mapping systems such as those provided by Churchill Navigation or CarteNav Solutions are now PC based. They offer augmented reality capabilities, whereby mapping data is blended with live video imagery to overlay the street and location data on the TFO's screen in real time.

While these electronic aids have greatly improved efficiency, paper maps and map books will always remain as a trusted backup option. As a TFO in training, one of the most critical skills to be developed is the ability to read a map, relate it to the world outside and then to navigate from A to B quickly and effectively day or night. This is done by using a Geographia or A–Z map book, which shows every motorway, road, residential street, waterway, park and other landmarks in clear detail. TFO selection and training focuses on an individual's ability to relate a moving 3D world to a static 2D map and to make accurate target identifications. This skill remains a key requirement regardless of technology, as there is always a need when working on task to be able to talk officers on the ground on to a target. The more a crew flies in a specific operational area, the more familiar they become with it and its landmarks. The onset of NPAS has taken city crews such as those at NPAS London out into the rural Home Counties and rural crew into the big cities, such as London, Birmingham, Manchester, Liverpool and Cardiff. The skills needed to navigate accurately in both an urban and rural setting are today more crucial than ever.

Skyforce Observer mapping

The EC145 is fitted with a Skyforce Observer Mk3 mapping processor located in the rear equipment rack alongside many of the other equipment control units. Mapping datasets are loaded, meaning that it is possible for each helicopter to have a mapping dataset with layers appropriate to their specific area of operation. The mapping processor is linked via a data link to the Wescam MX15 MCU, allowing the two systems to communicate with each other. GPS data is shared, and as such it is possible for the mapping system to send target coordinates to the Wescam MX15, allowing the camera to slew to target. It is also possible for the camera system to send a camera point of impact to the Observer mapping system and for it to be displayed on the mapping screen as a cross hair. This feature is exceptionally useful when following vehicles or managing pursuits, as the exact location and even the estimated speed of the target (as indicated by GPS) are displayed on screen for the rear seat TFO, the tactical commander, to use in their radio commentary.

Navigating around complex controlled airspace requires close cooperation with ATC. Obtaining clearance for police helicopters to fly to any given location in London is actually relatively simple, and relies upon a common reference source for guaranteed communication and understanding. Each page of the Geographia atlas for London covers a geographical area of approximately 40km^2 (5km × 8km). ATC references the page numbers as square numbers and gives clearance to the pilot to a specific square number. This indexing

BELOW The front of the Geographia map book shows the page numbers, which are converted to square numbers to aid communication with ATC.

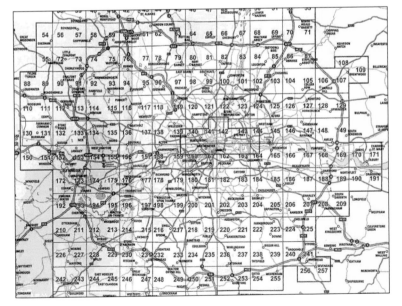

system is also used by the Met's command and control system known as CAD meaning that when a task comes into the ASU a page number is given along with an OS map reference. This page number is then communicated to ATC as a request to fly to a specific square number. Clearance is given to fly there, with any routing considerations required and a height below which the helicopter must operate. The trust that ATC have in the police pilots means that they can confidently identify the exact geographical area in which the helicopter is operating and can deconflict other air traffic.

The tactical commander enters the OS map reference to generate a target position along with a track direct to the scene, the distance from it and the estimated time of arrival. It is then possible to view the current position of the helicopter as it progresses along that track or to focus upon the target. The mapping coordinates of the target can also be sent to the Wescam MX15, and the front seat TFO can slew the camera on to the target. The mapping system has a number of layers that can be selected by zooming in and out, using buttons displayed on the screen. TFOs can zoom quickly in and out through the various mapping layers and change the mapping view from aircraft position to target position with ease. Mapping layers start with aviation charts at 1:500,000 and 1:250,000 and run through OS 1:50,000 before reaching the Geographia A–Z level, ending at street level with individual building outlines and house numbers clearly displayed. The variety of mapping views available enables the crew to navigate safely and effectively through controlled airspace using aviation mapping and then switch to tactical mapping when on task. If required, the TFO can zoom in to identify a specific address, location or house number.

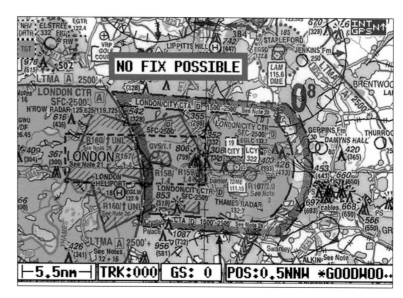

TOP A screen grab of the aviation mapping layer.

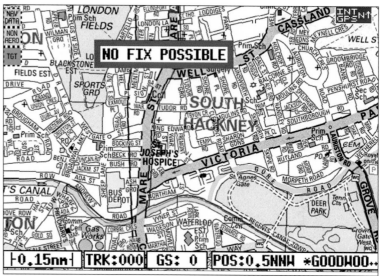

CENTRE A screen grab of the Geographia mapping layer.

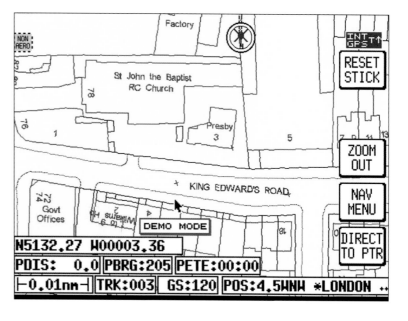

RIGHT A screen grab of the building outline mapping layer.

this would typically be the input from either the colour daylight TV camera or the IR picture.

Display technology has advanced so far that the units are now smaller and lighter yet deliver a larger and brighter viewing area. In 2006 a mission display system was launched that could interface multiple displays and video recorders with minimal wiring, and afford the crew a wide range of functionality and control from each and every display. The Skyquest (now Curtis Wright) developed VMS allows crew members to view and control the video imagery in the cockpit. The latest upgrade of the NPAS EC135T2 helicopter has seen this advance even further, as displays now have embedded computers and can run sophisticated mission software. The Carte Nav mission and mapping system that is installed has been customised to work seamlessly with the Avalex mission displays. This presents the TFOs with a huge range of functions and operational capabilities for imagery and mission management.

The Skyquest VMS manages video distribution in the EC145 cockpit through four TFO mission displays. The fifth display is a 6in display developed for the pilot; it has touchscreen control and eight function keys to allow the pilot to select available video inputs. The mission displays enable the TFOs to view all the available video imagery on a single display

ABOVE A screen grab of a QUAD image, showing the link between the Wescam sensors and the mapping that is used to identify an address quickly and easily.

RIGHT The tactical commander's workstation in the upgraded EC135 helicopters gives a nod to previous designs by offering two large displays side by side. The workstation is hinged, allowing the left-hand display to be angled towards the tactical commander or a rear passenger for operational use. The HD Avalex displays with embedded computers work well with the Carte Nav mission software to display the imagery from the Wescam MX10.

Police mission displays and video management

Police helicopters have always relied on TFO mission displays to deliver the imagery from camera sensors into the cockpit. These have developed over the years in line with general display technology, moving from early colour CRT displays through a variety of LCD displays to reach the current high definition touchscreen displays. In early installations it was only possible to view one video source at a time;

LEFT The cockpit of the EC145 in flight from the tactical commander's position. Communication in the cockpit is by voice, as it is difficult for the crew members to see each other: clear verbal communication is essential.

using a multiplex system called a QUAD. The video display system unit contains a number of QUAD cards that are capable of taking any four available inputs and then outputting them as a single video image on screen. The QUAD card is controlled by touchscreen commands, and the TFO is able to quickly bring any of the quadrants up to full screen display by simply touching the display within that image. Returning to the QUAD view is simply a case of touching the display screen again.

Each of the mission displays has 32 function keys located around the display screen bezel with eight buttons on each side.

Top – The eight buttons at the top of the display are used to select and control the first of the helicopter's video recorders. The evidential recorder can be started and stopped and the replay recorder controlled for replay in flight. Green LEDs illuminate to indicate the recording state and flash as the recording memory runs low.

LEFT The rear of the EC145 from the Silver (Eagle One) seat. The fold-down additional display is in the foreground; the main workstation is operated by the tactical commander.

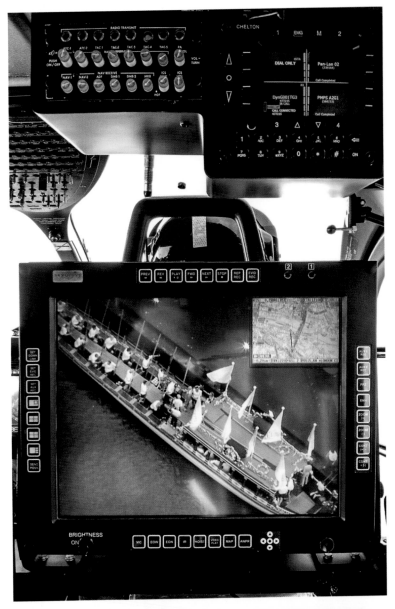

LEFT The VMS system allows the tactical commander to compose any imagery on screen and send this to the video recorder or digital video downlink. In this picture the radio control head is visible above the right-hand 15in display and a clean VIC image has been selected with embedded mapping imagery. This provides the control room with a nice clean image without the confusing symbology and a map to indicate where the camera is looking. In this case the subject is a ceremonial boat procession on the River Thames.

Right side – On the right of the display are a variety of function control buttons that deliver additional functions to the TFO. These include FRZ, which freezes the image on screen; ENH, which enhances an area at the centre of the image to see through mist and haze; and controls for the two auxiliary video recorders, allowing them to be started and stopped.

Bottom – On the bottom left of the display is a power and brightness rotary switch and on the bottom right there is a joystick controller, which is active whenever mapping is selected on screen and controls the mapping cursor for target selection. The eight function keys allow one-touch access to full screen imagery from the sensors available.

Left – The buttons on the left of the display give the TFO access to the various QUAD shapes. The default contents are preset, but pressing and holding the button enters program mode. The video source sent to the various video recorders and digital downlink is again set by default, although it can be changed by pressing and holding the SET buttons to the left of the display. The SEND POS button works in mapping mode to send the target selected by the TFO to the Wescam MX15 for camera slew to target.

LEFT The author as tactical commander on the EC145, operating the mapping screen joystick and showing the mapping keyboard used to enter target coordinates.

The pilot display operates separately from the other mission displays, with the pilot able to select any of the video sources available using the function keys along the bottom of the display. The display folds away for take-off and landing and gives the pilot the ability to select a different field of view to that used by the TFOs. A recent mid-life upgrade of the NPAS London EC145 fleet has seen these changed to Curtis Wright displays with generic buttons and the button legends appearing on screen. This has improved reliability and also ease of maintenance, as defective displays can now be easily swapped.

The EC145 is fitted with two video recorder units located towards the rear of the centre console; each of these has two video recorders, and it is therefore possible for the crew to record three separate video sources simultaneously on to removable flashcards. The fourth video recorder is a video loop, whereby the recorder records on to an internal memory the last few hours of evidential video recording. When the internal memory becomes full the recorder starts overwriting the earliest recording. The crew is able to stop, rewind and replay the video recorded by the evidential recorder in flight to review evidence. The video in the evidential and two auxiliary recorders is captured in MPEG-2 format at DVD quality on to removable flash cards. The video recorders available today and fitted to some NPAS helicopters record using MP4 format in high definition and are much more efficient than previously. In addition to recording video from the helicopter camera sensor, it is also possible to record audio from the police tactical radios (TAC1–4). The audio

LEFT The rear of the centre console in one of the recently upgraded NPAS H135 helicopters. This extended console presents all the mission equipment to the tactical commander for use in flight. Included are two Avalex digital video recorders that record in MP4 standard video. A button enables the TFO to mark an event on the video, and a video and USB interface panel enables the crew to upload data and video to the helicopter mission systems.

CENTRE The pilot's 6in display folds away cleanly for take-off and landing and provides the pilot with excellent situational video imagery or mapping.

RIGHT The NPAS EC145 fleet has recently undergone a limited mid-life upgrade of the mission displays with new Curtis Wright displays fitted and interfaced with the VMS. The hard function keys surrounding the display have been largely replaced by on-screen soft buttons.

source is selected by the front-seat TFO, using a rotary switch on the audio record selector panel in the centre console.

Airborne data link – downlinking of live video imagery

One of the most operationally useful capabilities of a police aircraft is the ability to broadcast imagery from the onboard camera sensors to the ground for the situational awareness of command teams and control rooms. A Home Office-led airborne data link (ADL) project delivered a national interoperability capability, with all 43 forces in the UK equipped to receive encrypted imagery from any of the UK police helicopter fleet. All NPAS helicopters and fixed-wing aircraft are fitted with a digital downlink capability using a new five channel plan and AES 256BIT encryption standard. This means that NPAS aircraft can fly anywhere in England and Wales and, by selecting the correct broadcast channel and encryption key, deliver live imagery into a forces command and control set-up.

The ability to downlink high-quality pictures reliably from a police helicopter was always seen as an operational priority, and the equipment needed to be relatively simple for the crew to manage in flight. The transmitter was installed alongside the power amplifier in the rear equipment rack, a control panel installed in the centre console and an actuator mounted antenna on one of the helicopter's steps.

The antenna

To improve performance on skidded helicopters, the decision was taken to install the transmission antenna (typically the size of a small tin can) on the end of a retractable arm known as an

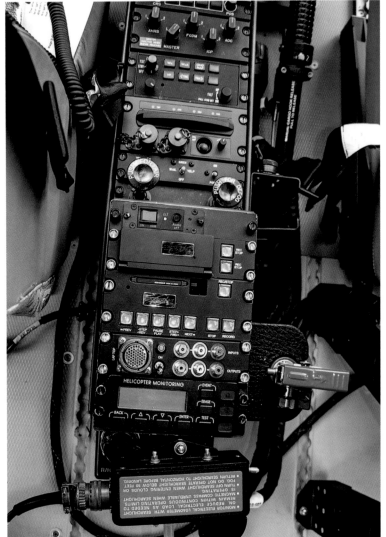

LEFT The rear of the centre console of the EC145 showing the two video recorders. One has two recorders, giving the ability to record evidentially on one using a removable flash card while the other replays the recorded video in the cockpit. The second recorder has two removable flash cards and is referred to as the auxiliary (AUX) recorder.

RIGHT The digital video downlink actuator fitted to the EC145 fleet. The actuator is lowered via a switch on the downlink control panel when the helicopter is in flight. This puts the transmit antenna below the skid for an uninterrupted transmission. An annunciator on the helicopter instrument slant panel indicates the status of the actuator.

actuator. A fixed position antenna attached to the belly of the helicopter or the cross beam was considered, but tests demonstrated that this induced a greater level of error in the received transmission and impacted upon receiver reliability and picture quality. The transmission pattern of the antenna was described as being similar to a ring doughnut shape, with the antenna being in the centre. As the helicopter banks and turns, the skids (solid pieces of metal) break into the doughnut-shaped signal and interrupt its passage to the receiver site. Moving the antenna below the skids once the helicopter is in flight reduces or even removes this risk, creating a reliable transmission from the antenna that gets to the receiver site regardless of the helicopter position or attitude.

Actuator

The actuator is lowered or retracted after take-off and in preparation for landing, which keeps the antenna clear of the skids and allows for a normal, safe and unobstructed landing. The raising of the antenna is a safety critical function and typically takes place as the helicopter prepares to land; the airspeed is lowered and the pilot gives permission to

CENTRE The front TFO as the EC145 prepares for landing. On the screen it is possible to see that the downlink actuator has been pictured so the pilot can monitor its safe and correct retraction before landing.

RIGHT The EC145 pictured in a hover over London. The digital video downlink actuator and antenna are clearly visible below the level of the skids. This ensures good-quality video transmissions during helicopter orbiting or manoeuvring.

system required the development of a remote control panel that would fit into the centre console of the helicopter using DZUS fasteners. The ADL project saw a mid-life upgrade of this panel and the downlink transmitter as the new national channel plan and encryption standards were implemented. The updated panel replicated the LCD display on the front of the transmitter; it was now possible for the crew to control every aspect of the downlink and uplink receiver system from a single control panel.

Ground receive infrastructure

The downlinked pictures are used on the ground by a multitude of policing and other resources. A ground infrastructure takes the received imagery and distributes it securely to the various police control rooms. Their ability to see exactly what the helicopter crew can see adds significant value when it comes to resource deployments, tactical decisions and situational awareness. Helicopter imagery is frequently the centre of attention within the control room and is displayed on large video display screens. At

ABOVE The upgraded digital video downlink control head, which gives the crew full control over the encryption, transmission and channel as well as the control of the antenna actuator. This was fitted after the national airborne data link project provided a common system to every police helicopter and force, giving a truly national transmit and receive capability.

raise the antenna. This is then monitored by the front seat TFO, using the camera system to see that the antenna is safely in the up position – something that is confirmed by the slant panel annunciator.

Control panel

The initial installation of the digital downlink

RIGHT The air support pod within SOR. This is resourced by an air support supervisor as tactical adviser to the command team. The helicopters deployed on the operation are managed using a range of technology to control radio, video and computer communications.

LEFT The video downlinked from police helicopters is not only viewed in police control rooms but is also shared with other emergency services and partners such as fire brigades, ambulance services, the coastguard and traffic control. In this picture a London Fire Brigade Command Unit is training to use its technology to view police helicopter imagery. *(Vislink)*

a more local level, portable receivers enable ground resources at an incident to view imagery directly from a helicopter that is working overhead. This capability is shared with partners from other agencies, such as the Fire Brigade and Ambulance Service, British Transport Police and Coastguard. At major incidents imagery from the police helicopter is considered crucial for the command teams of all emergency service organisations. Technology also enables this imagery to be distributed via the 4G LTE network to phones, computers and tablets on a secure network; this uses technology similar to that employed by the many commercial video streaming services. There is no doubting the value of air support, and it is best realised when the live imagery is distributed as widely as possible, allowing the maximum operational effect.

Uplinked imagery

It was always of the intention that the ASU would be able to view imagery from the ground. This proves useful when handing over from one helicopter to another at a protracted incident. The EC145 has a diversity receiver fitted within its rear equipment rack, and this interfaces with the Skyquest VMS and two receive antennas mounted on the helicopter cross beams. It is possible for a transmitter located in central London to broadcast any imagery from the London CCTV network, and for those images to be viewed by the helicopter crew. A specific installation was designed and approved: this

was lighter, but retained the uplink receiver and mapping processor. Imagery was displayed on a built-in 10in display, which has become known as the mini rack. This enables the retention of mapping and uplink when the helicopter is in special roles configuration.

Police searchlight

With early camera systems the IR sensor was not of sufficient quality to allow the crew to accurately identify heat sources; it was common to use the searchlight, often referred to by crews as the Nightsun, to provide light so the daylight camera can see what the heat source was. In the past, many a sleeping urban fox suddenly found itself in the centre of a police helicopter spotlight. Current IR sensors are of such high resolution that it is now possible to identify the

BELOW The two video uplink diversity antennas fitted to the rear cross beam of the EC145. Both antennas receive the transmitted video signals and the diversity receiver unit combines them to ensure a perfectly received video image.

source of the heat without the need for added light, so the use of searchlights has diminished. Tactically, however, they are still used to illuminate the ground, allowing ground resources to work safely and illuminating those on the ground during public order infringements, to show participants that they are being watched and that evidence is being collected. In night-time ad-hoc landings the searchlight is also used prior to landing to allow a visual recce and also to illuminate the landing site. While the searchlight may not be considered the most crucial piece of police role equipment today, it is still exceptionally useful – and most crews would rather have one available than not.

On the EC145, the SX16 searchlight is mounted on a bolster on the front of the port-side step, in a similar fashion to the Wescam MX15 on the starboard side. The front seat TFO operates the camera system and the rear TFO operates the searchlight. The range of movement possible for the SX16 gimbal presented problems, as it transpired that in order to prevent light coming into the cockpit via the chin bubble the software and hardware driving the gimbal had been limited to permit a range of movement between 90° to the left of the helicopter and 15° to the right of the nose. This meant that it was not going to be possible to fly a right-hand orbit with the pilot and rear TFO looking out of the helicopter windows and controlling the searchlight. The only way that the searchlight and the camera sensor would be coordinated was to fly a left-hand orbit with everything out of the left-hand side of the helicopter. This

BELOW The Spectrolabs SX16 fitted to the EC145, with its IR filter in place. This allows the illumination of the ground in IR or 'black light'. It was never used in the UK, though, so was later removed from the searchlight.

severely limited the pilot's field of view, so the crew had two options: either the searchlight had to be controlled by the front left TFO and the camera from the back seat or the rear seat TFO had to be unstrapped and work out of the left window; but both options were operationally limiting.

The second challenge came in respect of the way in which the SX16 was powered on and off, using a switch in the cockpit overhead switch console. When the EC145 entered service there was a potential safety issue in conversion training regarding switching on the searchlight. Switching on the SX16 meant identifying the correct switch in the roof console and moving it to the ON position, before moving it further to IGN and holding it until the lamp illuminated. This spring-loaded switch sprang back to the ON position when it was released. The roof console switches are backlit and clearly identified, but to read their captions the crew need to look up and backwards.

During TFO training the pilot gave permission for the front seat TFO to switch on the SX16. The TFO reached up into the roof switch panel, and moved the switch forwards through ON to the IGN position, expecting the light to buzz and illuminate. The searchlight was stowed looking forwards, and therefore when it illuminated a beam of light would easily be seen; but nothing happened. At this exact moment the pilot reported a sudden loss of hydraulic power, which quickly rectified itself. The crew discussed the situation briefly and decided to try again; the result was identical. It was only at this point that the switch the TFO was operating was checked. It transpired that two rows away from the SX16 switch, but in an identical position, was the hydraulics test switch – which allows the pilot to remove hydraulic power from one system while checking the other. The training sortie was abandoned and a debrief was completed, which looked at all the crew resource management (CRM) and other issues. It was concluded that the operation of switches by TFOs in a pilot-controlled area of the cockpit presented a challenge, so a process was soon developed that involved challenge, checking and positive confirmation. This ensured that the correct switch had been

RIGHT The power on switches for the SX16 are located in the roof switch console of the EC145, making the illumination of the light slightly more complex for a police crew. The proximity of the switches, which are always operated in the dark, to the hydraulics test switches caused a minor problem in training. This was rectified by a clear SOP.

FAR RIGHT The SX16 hand controller is very simple, with a joystick to control azimuth and elevation movement and a focus button that scrolls the lamp through from narrow to wide spot and back to narrow again as it is pressed. The power on switch is inactive; the crews were used to powering on the light from the hand controller in previous installations.

identified by the TFO and was only operated on the pilot's command.

Control of the SX16 is simple: a joystick is used to direct movement in elevation and azimuth, with a focus button being pressed to cycle through the searchlight shape from a large to a narrow spot. The hand controller is stowed at the rear of the centre console and as such is easily accessible by the front and rear crews. The pilot can also operate the searchlight using slave controls present on the collective pitch lever. Despite the limitations of the installation, the SX16 remains in regular use and is carried as a standard piece of role equipment.

CENTRE NPAS EC135 G-POLF was formerly G-ESEX and has recently been upgraded by Babcock to include a Spectrolabs Nightsun II searchlight on a bolster mount at the rear of the starboard step. This installation puts the searchlight and camera sensor on the same side of the helicopter. This is the installation that NPAS London EC145 crews wish had been fitted to the EC145 fleet.

RIGHT The Spectrolabs Nightsun II provides upgraded optics and bulb to give a clean and effective ground illumination. These searchlights are standard across most of the NPAS fleet.

Chapter Four

Police helicopter crew

A police helicopter with state-of-the-art role equipment is only as good as the crew members who operate it. Police pilots are instrument rated with high flying hours experience under their belt, while TFOs are exceptional police officers who bring their policing skills and knowledge into the air. When the pilot, TFOs and helicopter work effectively together the results are spectacular.

OPPOSITE Air support crews operate 24 hours a day and frequently witness the natural beauty of flying as day transitions to night. Here the cockpit instruments of the EC145 glow against the night sky at sunset over London. *(John Roberts)*

Police helicopters in the UK are typically operated by a crew of three, comprising a pilot and two TFOs. Elsewhere in the world police officers are trained within their units to fly helicopters, which are typically operated in a dual-pilot configuration. In the UK the CAA determined fairly early on in the development of police aviation that pilots should be experienced pilots and TFOs police officers. The two roles and their requisite qualifications, experience and requirements were separated and have remained so. The police helicopter is operated by the crew, although in technical terms the police officer TFOs are actually classified as passengers – with the helicopters operated to public transport safety standards. It appears that in the development phase of the PAOM chief constables had a view on how their TFOs should be treated.

Police officers are appointed and are not employees, and as such many of the standard employment conventions do not necessarily apply to them. They are protected by Police Regulations, which outline what a chief officer can and cannot do with them and their hours, places and conditions of work. Being a police officer is a great profession and conditions of

work are generally good, but crime, disorder and other policing issues are not always predictable. Chief constables therefore need the ability to deploy police resources as, when and where they are needed. If TFOs had been classed as crew it would have invariably meant that they would have needed to follow an agreed flight time limitation. This, similar to that used by aviation professionals such as cabin crew, would set out how many hours they could fly and how much rest they needed in between duties. They would not be contactable on days off and working into rest periods would have a significant impact, should retention on duty or recall to duty be justified in order to fulfil a policing function. Police chiefs therefore determined that they should not be crew, and that to retain their function as warranted police officers they would be classed as passengers.

Pilot

Outside military helicopter flying there are few civilian roles that are as dynamic and exciting as police helicopter operations. At Lippitts Hill almost all the police pilots entered air support following military service, having

BELOW Ground handling the helicopter using the TLC heli-lift. This hydraulic lift is electrically powered and uses clamps on the extending arms to pick up and secure the helicopter, which can then be easily manoeuvred on the apron. Procedures dictate that two crew are always engaged in ground handling: one drives the heli lift and the other checks for clearance and safety.

amassed many thousands of flying hours in the Royal Navy, flying Sea King or other search and rescue (SAR) helicopters, in the Royal Air Force flying Chinook helicopters, or in the Army Air Corps. The non-military pilots have come from a variety of commercial flying backgrounds, with their common ground being their level of flying experience. A police pilot must hold a commercial pilot's licence (helicopter) with an RT licence and a Class 1 medical (unrestricted), as well as an EASA single-pilot instrument rating. A minimum of 1,500 hours' total helicopter flying time which must include 50 hours flying at night (20 hours as pilot in command), 500 hours as pilot in command over land under visual meteorological conditions and significant low-flying experience. The minimum requirements reflect the complexity and demands of the police flying role. This requires a pilot who with minimal planning can take off and undertake whatever mission is presented, thinking dynamically and staying safe in some of the most complex airspace in the world. There is little time for flight planning: crews brief at the start of a shift in relation to the weather, airspace issues, aircraft performance, load and emergencies, then prepare their helicopter and equipment for an immediate lift should the call come through. In the days before NPAS pilots would become very familiar with their normal operating area and the airspace issues, limitations and hazards that it presented. Under NPAS pilots have a wider reach and are expected to be familiar with a much greater area of operation.

The pilot role involves the safe and effective operation of the helicopter and its flight systems. In order to do this pilots undergo regular operator proficiency checks (OPC) and instrument flight procedures (IFP) checks: these are done in a non-operational police helicopter with dual controls fitted and with an experienced type rated examiner. During training and assessment pilots are expected to demonstrate mastery of all the helicopter's systems, to be able to fly safely and deal with a wide range of emergency situations, including autorotational landing in the event of double engine failure. The training and assessment also includes a ground-based examination, with pilots expected to deliver a faultless performance: standards are necessarily high and frequently exceed the

requirements set by legislation. It is this attention to detail and demand for the highest standards that provides the operator with confidence that whatever challenge is presented to police pilots, they are more than capable of undertaking the role safely and effectively.

The role of pilots may seem quite obvious: they fly the helicopter. It is common banter to joke about the pilots being 'drivers', and to remind them that the helicopter says 'POLICE' on the side and not 'PILOT'. However, they are truly an essential part of the crew. There are no egos in police flying and no pecking order when it comes to the cockpit: the pilot is the captain of the helicopter and responsible for its safe flight and operation; the crew, though, have an important role to play in supporting the pilot to achieve this – so they always brief, train and debrief together. At the start of a shift the pilot will receive a handover from the previous pilot, covering any issues with the helicopter: these could be defects or problems with specific items of equipment, hazards in the airspace that have developed recently or anything that the new pilot needs to know. The helicopter has an agreed minimum equipment list (MEL), which sets out everything that must be operational to fly. If a landing light bulb were to go, for example, what has to be done? Does the pilot ground the helicopter or can the defect be deferred? The MEL is comprehensive and is agreed by the maintenance organisation, the CAA and the operator to ensure that everyone is clear about the state of the helicopter and when it can and cannot be operated.

The pilot must be given at least 30 minutes of pre-flight briefing and preparation time at the start of the 12-hour flying shift. This involves obtaining and briefing themselves on the latest weather actuals and forecast for the area of operation, the Notices to Airmen (NOTAM) that relate to their area and period of operation and notify all pilots of anything impacting upon safety in their airspace that they should be aware of. NOTAMs can include captive advertising balloons, planned media or filming flights, air displays, defective navigational aids, closed runways and cranes with defective lighting. In summary, it is a general and unrestricted notification that is publicly available to everyone and aims to ensure that all pilots are fully briefed

RIGHT Passenger
safety briefing card,
p. 1. TFOs and
pilots undergo an
annual emergency
and survival training
course, which means
they can fly legally
without the need for
a full safety briefing
each time they
take off. All other
passengers must be
briefed by a member
of the crew (TFO
or pilot) and have
access to a laminated
passenger safety
card. This is exactly
the same as on any
commercial airline
flight.

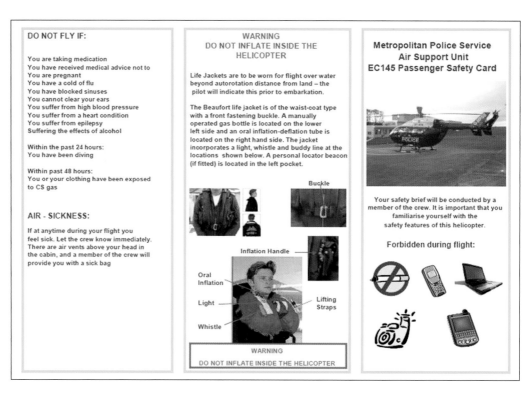

DO NOT FLY IF:

You are taking medication
You have received medical advice not to
You are pregnant
You have a cold of flu
You have blocked sinuses
You cannot clear your ears
You suffer from high blood pressure
You suffer from a heart condition
You suffer from epilepsy
Suffering the effects of alcohol

Within the past 24 hours:
You have been diving

Within past 48 hours:
You or your clothing have been exposed
to CS gas

AIR - SICKNESS:

If at anytime during your flight you
feel sick. Let the crew know immediately.
There are air vents above your head in
the cabin, and a member of the crew will
provide you with a sick bag

WARNING
DO NOT INFLATE INSIDE THE
HELICOPTER

Life Jackets are to be worn for flight over water
beyond autorotation distance from land – the
pilot will indicate this prior to embarkation.

The Beaufort life jacket is of the waist-coat type
with a front fastening buckle. A manually
operated gas bottle is located on the lower
left side and an oral inflation-deflation tube is
located on the right hand side. The jacket
incorporates a light, whistle and buddy line at the
locations shown below. A personal locator beacon
(if fitted) is located in the left pocket.

Buckle

Inflation Handle

Oral
Inflation

Light

Whistle

Lifting
Straps

WARNING
DO NOT INFLATE INSIDE THE HELICOPTER

Metropolitan Police Service
Air Support Unit
EC145 Passenger Safety Card

Your safety brief will be conducted by a
member of the crew. It is important that you
familiarise yourself with the
safety features of this helicopter.

Forbidden during flight:

with respect to anything that may affect the safety of their flight. Reading NOTAMs and understanding their content is, however, the personal responsibility for the individual pilot – and a professional pilot will always do this.

The helicopter's technical log (or more commonly the tech log) must be carried in the helicopter at all times and is used to record everything that happens to that helicopter: maintenance inspections, repairs, flights, engine starts, defects and so on. The majority of component parts on a helicopter have a maintenance regime attached to them, either with a timed lifespan (e.g. 1,000, 2,000, 5,000 flying hours) or an inspection interval (50 hours, 100 hours, etc.), at which point they are replaced. It is therefore essential that a log of the flying hours for each individual helicopter is always available to inform correct maintenance schedules and any specific serviceability issues. This ensures safe and legal operation at all times.

Every helicopter needs a daily maintenance inspection that has a currency of up to 24 hours. This is referred to as a Check A and must be completed every single day without exception. The regulations state that when the helicopter is operated from a maintenance base the engineering team should undertake this, which can take up to an hour to complete. However, the role can be delegated to a pilot with suitable training and approval, and as such the pilots may well undertake Check A as part of their daily role. Operationally the helicopter is technically off line while it is taking place, and its timing is controlled. This ensures first that the helicopter does not need to be grounded in the middle of an essential tasking and second that it is undertaken at the traditionally quieter times of the day. In summary, the pilot needs to know the operational status of their helicopter airframe and how many flying hours or what time period remains until either maintenance must be undertaken or Check A expires and must be revalidated.

Having briefed themselves ready for the shift, received the helicopter from the outgoing pilot, completed the handover, signed the technical log and completed a walk-round visual inspection, the pilot declares the helicopter ready to go on line, and leads the briefing of the crew for the day's policing activities.

Tactical flight officers

Police helicopters in the UK are typically crewed by two TFOs in addition to the pilot. TFOs are usually police officers of constable or sergeant rank, although in the

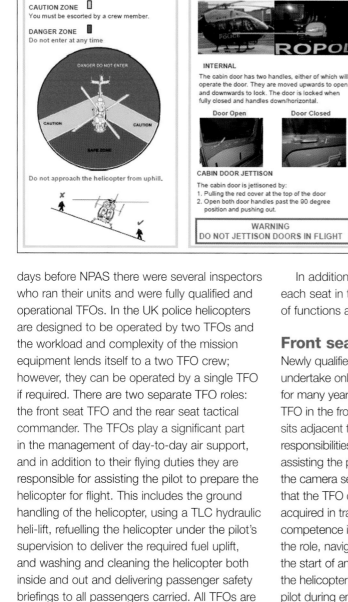

HELICOPTER APPROACH & DANGER AREAS

Always approach the aircraft under the direct control of a crew member or through direct visual contact with the pilot in the right hand seat.

SAFE ZONE
Approval from pilot by thumbs up hand signal or steady landing light and boarding light.

CAUTION ZONE
You must be escorted by a crew member.

DANGER ZONE
Do not enter at any time

DANGER DO NOT ENTER

CAUTION   CAUTION

SAFE ZONE

Do not approach the helicopter from uphill.

DOORS & EMERGENCY EXITS

EXTERNAL
Cockpit door and sliding cabin door handles Are operated upward to open and down to close. The cockpit doors give a green/red indicator strip. At the bottom to indicate the door is fully closed. The cabin door is of the 'plug' type and must be moved outboard to enable it to slide open.

Door handle   Cabin Door handle

INTERNAL
The cabin door has two handles, either of which will operate the door. They are moved upwards to open and downwards to lock. The door is locked when fully closed and handles down/horizontal.

Door Open   Door Closed

CABIN DOOR JETTISON
The cabin door is jettisoned by:
1. Pulling the red cover at the top of the door
2. Open both door handles past the 90 degree position and pushing out.

WARNING
DO NOT JETTISON DOORS IN FLIGHT

HARNESSES

The three passenger seats are situated in a fixed position, towards the rear of the aircraft. A crew member will show to the appropriate seat. Each seat is fitted with a fully adjustable four point harness. Simply push all four straps into the buckle slots and adjust straps to a comfortable position

To release the harness simply twist the buckle in an anti clockwise direction, as indicated.

EMERGENCIES

In the unlikely event of an emergency arising, The pilot will endeavour to keep you informed of the situation. Follow any instructions given.

Do NOT try to talk to the pilot
Do NOT release your seat belt or open the doors until the aircraft and rotor blades have stopped.

If you have any questions regarding this briefing. Please ask a member of the crew.

LEFT Passenger safety briefing card, p. 2.

days before NPAS there were several inspectors who ran their units and were fully qualified and operational TFOs. In the UK police helicopters are designed to be operated by two TFOs and the workload and complexity of the mission equipment lends itself to a two TFO crew; however, they can be operated by a single TFO if required. There are two separate TFO roles: the front seat TFO and the rear seat tactical commander. The TFOs play a significant part in the management of day-to-day air support, and in addition to their flying duties they are responsible for assisting the pilot to prepare the helicopter for flight. This includes the ground handling of the helicopter, using a TLC hydraulic heli-lift, refuelling the helicopter under the pilot's supervision to deliver the required fuel uplift, and washing and cleaning the helicopter both inside and out and delivering passenger safety briefings to all passengers carried. All TFOs are specifically trained and authorised in these roles and without them the ASU would not function correctly. It is also important to remember that just like any other police officers the TFOs are required to produce the evidential paperwork to support the video or photographic evidence that they routinely deliver. It is for these reasons that TFOs and pilots are kept busy on the ground between flights.

In addition to the generic TFO functions, each seat in the helicopter has a different range of functions and responsibilities.

Front seat TFO (Obs 1)

Newly qualified TFOs are initially trained to undertake only the front seat TFO role, known for many years as Observer One or Obs 1. The TFO in the front left-hand seat of the helicopter sits adjacent to the pilot and has a number of responsibilities; these include supporting and assisting the pilot, navigation and operation of the camera sensor. It is in this front seat role that the TFO can consolidate the knowledge acquired in training and gain confidence and competence in relation to two essential parts of the role, navigation and camera operation. At the start of any mission, when the crew get into the helicopter the front seat TFO supports the pilot during engine start by undertaking a safety look out and maintaining eye contact with the tactical commander; the full start-up procedure will be examined in detail separately. They then apply power to the police mission equipment as soon as the avionics master switches have been powered on by the pilot and permission for role equipment power on has been granted. They monitor the power on routine of the VMS, camera sensor and police tactical radios, and generally

prepare the equipment for use. This power on routine has been carefully developed to ensure that the mission equipment is switched on in the correct predefined sequence, allowing it to interface correctly with other items of equipment. This is essential, as if this doesn't take place the GPS-assisted auto camera mode won't work and the link to the moving map will fail. The crew rely heavily upon the mission equipment working correctly, and while the mission can be undertaken with the camera in manual mode and without the moving map, things are more complicated – and when the crew are under pressure the work rate goes up considerably.

The power on routine starts at the back of the centre console:

1 **Check video recorders have flash cards. Press VMS switch to apply power. Press MX15 switch to apply power.**

2 **Switch on Skyforce Observer at the control head** – this applies power to the mapping system processor.

3 **Check that the audio record switch is set to TAC1** – this means that as a default the audio recorded on to the video will be from the main working radio, typically TAC1. This can be changed during the flight as required, but seldom is.

4 **Check the CH450 control head** on the slant panel to see that the police tactical radios are booting correctly. Check individual station box audio and radio settings.

5 **Apply power to the Wescam system via the hand controller. Monitor power-up and set-up camera preferences. STOW camera for take-off.**

6 **Monitor the power-up on the front TFO display** – the screen can be used to check all video inputs and camera settings are correct.

The power on routine can be monitored from the display on the front seat TFO mission screen. Powering on the camera sensor and checking its operational control in azimuth and elevation

BK117 C-2 EN

EC145 MEL

ABOVE The instrument panel of the EC145 helicopter showing the flight reference cards and minimum equipment list, stowed on the top of the panel for easy pilot or crew access.

is important, along with zoom and focus control; these are the critical features that must always work correctly. The IR sensor takes a few minutes to cool down to operating temperature, and as such remains unusable for the first few minutes of flight. This only causes a problem if the task location is really close to the base.

RIGHT The etched checklists showing pre-take-off, post-take-off and pre-landing checks. This plate is attached to the instrument panel and the list is replicated on the front TFO's fold-out map chart holder.

Pre Take Off	After Take Off	Pre Landing
Performance	Strobes	Performance
CAT A Switch	CAT A Switch	CAT A Switch
Fuel / FLI / Ts & Ps	Altimeters	Landing Lights
Warning Unit / CAD	Landing Lights	Altimeters
Standby Horizon	Weather Radar	Screens
TCAS	Downlink	Strobes
Altimeters		Downlink
Radios / Nav Aids		Doors / Harnesses
Controls		
Landing Lights		
Doors / Harnesses		

Having completed the power on and checking of police mission equipment, the front seat TFO assists the pilot with the pre-take-off checks. There are three routine checklists in addition to those contained on the flight reference cards (FRCs) in the emergency and routine sections. The FRCs are used in the event of an in-flight warning caption or emergency, and while the pilots know from memory the checks and actions they should take, the front seat TFO takes out the FRCs, finds the appropriate page and illuminates it so the pilot can reference and cross-check it. Thankfully this is not a common occurrence in operational flights, although in-flight emergencies such as chip warning lights or other system issues or failures do happen from time to time. As a routine, the TFO runs through a checklist, requiring a verbal response from the pilot; the front seat TFO checks and challenges this as appropriate. These checklists are completed pre-take-off, post-take-off and pre-landing, and are common to each and every form of aviation. The exact content and sequencing of the various checklists varies according to the type of aircraft being flown. On the EC145 they are etched on to a placard on the instrument panel and also appear on a laminated card stored in the front seat TFO's chart holder. The checklists were as follows:

PRE-TAKE-OFF CHECKS

Challenge	Response
Performance	Brief
CAT A switch	On
Fuel/FLI/Ts and Ps	Noted/in limits
Warning unit/CAD	Clear
Stand by horizon	Uncaged
TCAS	On
Altimeters	Set/TDP
Radios/navaids	Set
Controls	Unlocked
Landing lights	On
Doors/harnesses	Secure

AFTER TAKE-OFF

Strobes	On
CAT A switch	Normal
Altimeters	Set
Landing lights	Off
Weather radar	On

LANDING CHECKS

Performance	Brief
CAT A switch	On
Downlink	Up
Landing lights	ON
Altimeters	Set
Screens	Stowed
Strobes	Off
Harnesses	Secure
Weather radar	Standby

The check and challenge process mentioned above is critically important: it is easy to run through a checklist with a memorised response that does not match reality. For example, when checking 'Warning unit/CAD', the pilot should look at the warning unit and CAD and note their content. On most occasions the warning unit/CAD will be clear, but there may be a caption that is OK and expected to be there; this should be checked and acknowledged. Although police helicopters are a single-pilot operation, this is one example of the safety critical support role that the TFOs have.

Navigation

Having taken off and climbed to a safe altitude and speed, the after take-off checklist is completed – and the front seat TFO then focuses upon navigation to the task. Navigation for a police TFO differs from that of a pilot, in that the TFO is more focused on where the helicopter is going and finding the target and less on where the helicopter is and any airspace issues. However, to know where you are going and how long it will take to get there, you need to know where you are. In rural areas navigation may be based upon a compass heading from the current position and the monitoring of speed to calculate distance travelled and the picking out of visible features, such as major roads, villages, waterways, railways and water features. In an urban setting the navigation may be more focused upon recognition of local landmarks and features such as a specific tower block, water feature, railway junction, gasometer, park or building in daylight and light features at night. In daylight a park may have a distinctive appearance in terms of its shape and the presence of water features such as ponds; in darkness the park may be visible because of the shape of its darkness against the street lighting that is around it.

Even today TFOs navigate using a map book. They start by looking for large features to guide them towards a target, then medium-sized features and on to the smaller features as they become visible. If flying north to south across London, for example, the target heading may be identified relative to a large feature, such as the Shard, London Eye or Canary Wharf. The TFO initially instructs the pilot to head to the right of the Shard while working out the next visible feature. Navigation is done by shape recognition in daylight or darkness, with the visual picture differing greatly between the two. Over time the TFOs build up a strong mental picture of their operating area and use that knowledge to find themselves a starting point from which they can begin to look at the detail. In Whetstone, for example, there are two large tower blocks close to each other on the High Road, while to the south in North Finchley there is another tower block at Tally Ho Corner. From the ground these buildings just blend into the surroundings if you drive past, but when viewed from the air they tower above the other buildings in the area and stand

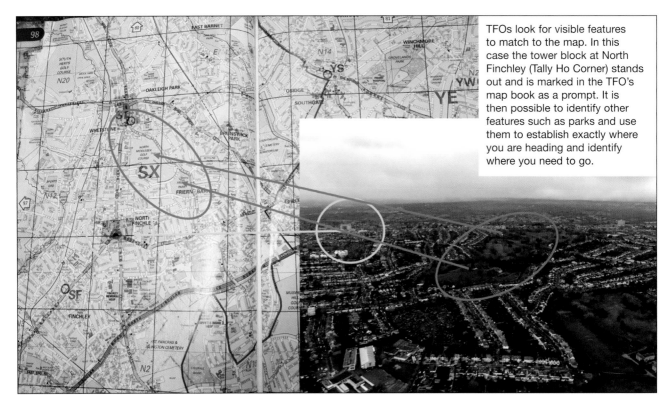

TFOs look for visible features to match to the map. In this case the tower block at North Finchley (Tally Ho Corner) stands out and is marked in the TFO's map book as a prompt. It is then possible to identify other features such as parks and use them to establish exactly where you are heading and identify where you need to go.

ABOVE An explanation of how map books are used to aid navigation to targets by identification of large visual features.

out clearly. To complete a task in this part of London, the TFO will look at the image these three buildings create and their relative position to each other. This means a target area can be picked out from some distance simply by shape recognition. The map books used by TFOs are personal to them, as they highlight and mark features that stand out specifically for them as they navigate about in daylight or darkness. Features may be an unusually shaped building or the bright lights from a railway goods yard, for example. The notes and highlighted items on any given page give the TFO a series of prompts.

Having arrived in the proximity of the target, the shape of the roads and relative position of junctions, parks, roundabouts and other features enable quick and accurate identification of a given street. The technology takes care of most of this, and while the front seat TFO still undertakes the basic manual navigation, the tactical commander has entered the map reference within seconds of receiving the task and the moving map works out the heading, draws a track line to the target and calculates the estimated time of arrival on scene. The mapping technology is a real bonus and improves efficiency and accuracy greatly. The

ability to give a precise time of arrival ('We will be with you in 1 minute and 20 seconds') has often been used by TFOs to impress those on the ground who are blissfully unaware of the assistance provided by a mapping computer.

Camera operation

On arrival at the scene, the front seat TFO is primarily focused upon the operation of the camera sensors, daylight TV (EOW), spotter scope (EON) and IR. Communication with the pilot is crucial, so that the pilot can position the helicopter and give the TFO the best possible view from the sensors. If the crew is following a crowd on a protest march, for example, and downlinking live imagery for the command and control centre, keeping the protestors visible on screen is key to delivery of quality air support. This is more of an issue in built-up urban areas as the height and position of buildings can severely obstruct the view. The camera has multiple sensors, and although the front seat TFO may only be focusing on one at any given time, the VIC, it is important that the other sensors are focused and of usable quality as they may well be recorded and are almost certainly being viewed by the tactical commander.

Rear seat TFO (tactical commander)

After around six months of front seat flying experience a TFO is ready for the tactical command course, which equips them to sit in the back seat of the helicopter and assume tactical command. This means they determine what the crew do and how they do it, in other words coordinating each sortie. The pilot is the captain of the helicopter and in ultimate command of the flight and its safety, but the tactical commander is in charge of the policing mission. The tactical commander has a significant safety role during the start-up procedure – monitoring the engine start for signs of fire and checking doors and hatches are secure before take-off. In flight their primary role is one of command and control communications and they manage the four police radios, controlling them from the Chelton RH450 remote radio controller above the rear workstation. The mission is effectively managed from this position: the two large mission displays allow the tactical commander to see all the sensor outputs from the camera system and to control the video recorders, digital video downlink and moving map system.

Communications

From the initial tasking message, which might be something such as 'vehicle pursuit A406 towards Hanger Lane, page 119, MPSINTOP1', to arrival on scene, the tactical commander ensures that the entire crew is fully updated about the task. This short thread of information is sufficient to tell the crew the task type (vehicle pursuit), the location and direction of travel (A406 towards Hanger Lane), the page number (p. 119 needed for ATC clearance) and the radio talkgroup (MPSINTOP1) that is being used to manage the task. Radio communications are crucial: location and details can evolve as the helicopter is en route to the task location, and these need to be considered. The tactical commander selects the correct operational radio talk-groups and ensures that the radio control head is set up to manage the incident, establishing effective communication with local and force-wide resources. During take-off the cockpit is sterile and there are no unnecessary communications with the pilot, which means

that typically the tactical commander is the only one monitoring these radios and setting everything up at this time. As soon as the post-take-off checks are complete communication in the cockpit opens up. The first thing the tactical commander does is brief the pilot and front seat TFO on where they are going and what they are going to do. The tactical radio set-up is explained and the audio recording required is communicated, so that the front seat TFO can adjust any switching necessary. With all crew communications an instruction is given and acknowledged with a repetition, ensuring that clear and precise understanding of what is required takes place.

Navigation support

The rear workstation is fitted with a keyboard. One of the first things the tactical commander does is to input the map reference for the target into the mapping system. The police control room requesting air support typically provides an OS map reference, which is converted to six or eight digits to provide the exact target location within a few metres. The pilot is directed towards a specific page number as an initial steer, which gives them the opportunity to talk to ATC and obtain flight routing clearance. Occasionally ATC will instruct the pilot not to fly beyond a specific location, for example asking them to stay north of the A40 when operating near to Heathrow Airport. The quicker the tactical commander can enter the target details into the mapping computer and obtain an accurate heading to target, the quicker an estimated time of arrival can be calculated. The mapping display draws a line from the helicopter location to the target and indicates the heading of the helicopter with a separate dotted line; it is easy to bring the target bearing and helicopter heading together to ensure that the helicopter is travelling directly to the required location. Airspace information can also be displayed to support the pilot, although most have an intimate knowledge of the airspace they are operating in and don't need digital support.

The tactical commander can also send the target location to the camera system as a mapping coordinate by pressing the F9 key on the keyboard or the SEND POS button on the mapping display. As soon as the target

ENG OIL P	VAR NR	ENG OIL P
FUEL PRESS	RP1	FUEL PRESS
TWIST GRIP	F PUMP FWD	TWIST GRIP
HYD PRESS	F PUMP AFT	HYD PRESS
OVSP FAIL	DOORS	OVSP FAIL
GEN DISC	BAT DISCH	GEN DISCON
PITOT HTR	RP2	INVERTER
		PITOT HTR

OFF

SCROLL

SELECT

BRT+

BRT−

END   FUEL kg

FF1
0
Kg/h

400   47   54

FF2
0
Kg/h

An explanation of how to read the fuel gauge in flight to calculate flight endurance as part of mission management. *(Author's Collection)*

1   Main fuel supply showing 400kg
2   Fuel supply tank No 1 engine (47kgs)
3   Fuel supply tank No 2 engine (54kgs)

EC145 burns roughly 4 x kg per minute
Minimum landing allowance (MLA) by day = 20 mins, by night = 30 mins

Total fuel to tanks dry = 400+47+54 = 501
Minus MLA (day) = 501−80 = 421
421/4 = 105 mins available flight endurance
The helicopter therefore has 105 minutes (1hr 45mins) of flight endurance before it must be on the ground at the base or a refuel destination.

BELOW Front TFO display with mapping.

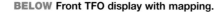

has been sent to the camera system the front seat TFO can choose the slew command from the camera hand controller, and the camera turret will automatically move (slew) to point at the target coordinates even if it is beyond visual sight. The system is so accurate that it is possible to select a specific house or building from the mapping system, send its coordinates to the camera sensor and slew the camera to that target. As the helicopter comes within visual range of the selected target it will appear on screen. The tactical commander can adjust the target on screen if necessary and send the new target to the camera system even if the scenario is dynamic and fast moving, a function that can be very useful.

The role of the tactical commander varies from task to task. During a vehicle pursuit the mapping system changes to show the camera position on the screen, meaning that the tactical commander can very accurately confirm exactly where the vehicle on screen is located, its direction of travel and an estimate, based upon GPS position, of its speed. This subtle change in the mapping system ensures that the technology is being used to maximum operational benefit and frees up the TFOs to concentrate on pursuit commentary support and ground resource coordination. The tactical commander controls everything the helicopter does on task, from ensuring that the correct imagery is downlinked to the control room or partner agency (fire, ambulance or coastguard) to monitoring the evidential video recording and managing all radio communications. In the cockpit the crew communicate constantly to ensure everyone is fully aware of what is going on, and that they are all actively contributing to flight safety and the effective completion of the task.

Fuel management

The tactical commander liaises with the control room to ensure that everyone is aware of emerging task requirements and advises the control room of progress. It is not uncommon for multiple tasks to be queued awaiting deployment, and each is assessed on the basis of its threat, harm and risk. The viability of a task may well be impacted upon by its location and the flight endurance remaining. The fuel state is displayed pictorially on the

CAD. Usage is monitored continuously by the pilot, who advises the crew of the remaining endurance throughout the flight, as well as the time that the helicopter must leave the task to allow sufficient time to return to base and land within MLA requirements. The EC145 can carry around 600kg of Jet-A1 fuel, with the exact load depending upon the weight of the crew and the number of passengers. In a police helicopter fuel load is often the only variable element of the payload within maximum all-up weight for flight.

The EC145 has two Turbomeca Arriel 1E2 engines that have a typical fuel burn rate of 4kg per minute between them (2kg each). Flight endurance is a simple calculation of the total fuel load carried, minus the MLA, which is 80kg by day and 120kg by night, divided by the burn rate of 4kg a minute. If, for example, the crew took off with 600kg of total fuel, the likely total flight endurance would be 150 minutes (600/4) minus the MLA of 80kg, which is 20 minutes; the total flight could be reasonably expected to be 2 hours 10 minutes (130 minutes). The tactical commander is responsible for ensuring that if a task is accepted there is sufficient flight endurance to get to the task location, complete the task effectively and return to base. For this reason, the digital fuel display in the CAD can be viewed and interpreted by the tactical commander throughout the flight, although fuel calculations are always checked and verified by the pilot. The crew typically know how much endurance they have on any given flight: they know what time they took off and as such know what time they expect to be landing. The monitoring of fuel displays and the calculation of remaining endurance is important, as they regularly need to adjust flight endurance based on actual fuel burn rate. Unlike the major airlines, which may be able to carry significant fuel reserves, police helicopters typically operate to maximum fuel load and use the available fuel while landing with the MLA required. The management of fuel is especially important when undertaking public order or security escort tasking, which usually involves a requirement to provide guaranteed air support cover over a specific period of time. This is achieved through careful planning of fuel endurance and taking into account how long the helicopter can fly before refuelling, ensuring take-off at an agreed

time and making sure that the helicopter is on task (or ground assigned) at an agreed time. If it is necessary to provide seamless coverage over a prolonged period of time then the coordination of two helicopters is necessary, with a handover from one crew to another on scene at an agreed time and location.

TFO responsibilities – start-up

Both the front and rear seat TFOs play a significant role during the start-up, pre-take-off and pre-landing phases of the flight. In order to ensure a safe and effective engine start, the tactical commander takes up a position outside the rotor disc to the front of the helicopter in clear sight of the pilot. This is for three reasons: first, to ensure that there are no personnel or hazards present on the apron during engine and rotor start; secondly to look for signs of fire during engine start; and thirdly to ensure that all doors and hatches are secure and ready for take-off. Communication between the crew members during engine start is typically done via agreed hand signals, with the signal given and acknowledged by relevant crew members. A remote intercom can be used, this enabling the tactical commander to plug into a wireless transmitter/receiver and obtain full intercom communications within their helmet even when outside the helicopter; however, this is always backed up with hand signals. These start with the pilot indicating the engine to be started; typically the first to start will alternate over various flights within a tour of duty. On an odd-numbered day (first, third, fifth, etc.) the first

BELOW Daily fuel testing is completed at every NPAS base to ensure that there is no water in the Jet A-1 fuel. Samples are taken from the helicopter and the fuel delivery system, and these are retained at the base.

RIGHT The tactical commander in front of the helicopter during start, monitoring the engine and rotor start. This safety start-up takes place day or night and in all weathers, including rain and snow.

flight of the tour will start engine 1, while on an even numbered day (second, fourth, sixth, etc.) engine 2 will be started. The second flight will start the other engine first.

After the pilot has indicated the engine that will be started first, the tactical commander moves towards the port or starboard side of the helicopter (engine 1 is port, engine 2 is starboard) so as to have a good view of the pilot, front seat TFO and the engine that is being started. One of the biggest risks in aviation is fire, and the ability to monitor engine starts to ensure they are free from fire is an essential component within a safety

RIGHT The tactical commander completes his safety walk around the helicopter, checking that doors and hatches are closed and secure ready for flight.

LEFT A TFO wearing full flight safety clothing and personal protective equipment (PPE).

1 **Alpha aircrew helmet** – provides protection against head injury, offers hearing protection and eye protection via clear and sunlight visors. Active noise reduction (ANR) delivers clear intercom and radio communications to the crew member via headphones and mike.

2 **Nomex flight suit** – the military Mk16 flight suit is made from fire-resistant Nomex fabric and is worn on top of long sleeved and long legged cotton base layers.

3 **Cotton undergarments** – cotton base layers are worn beneath the flying suit with the principle being that heat is dissipated via use of non-flammable layers. Non-cotton fabrics can melt in fire causing worse burns and are avoided.

4 **Soft leather flying gloves** – protect the hands and retain a tactile ability to feel controls and switches.

5 **Leather protective boots** – leather boots are worn to offer protection for the feet and ankles. These typically have a low profile pattern on the sole to avoid picking up stones and other foreign object debris (FOD) and depositing it in the cockpit where it could become a hazard.

6 **Tactical vest** – used to provide easy access to policing personal protective equipment (PPE) such as handcuffs and friction batons. CS spray is not carried by air crews.

RIGHT Monitoring an engine and rotor start when not at the main helicopter base presents additional challenges. Police helicopters do on occasion land in public places such as parks and the presence of members of the public around the helicopter can present safety risks. In these circumstances the tactical commander must keep a wider field of view to ensure that no one approaches the helicopter during start-up.

BELOW The tactical commander outside the rotor disc for engine and rotor start.
(Hugh Dalton)

management system. Having monitored the start of both engines, the tactical commander, with the permission of the pilot, enters the rotor disc. This positive command is typically a thumbs up. Crews are trained from the start of their air support career that they must always approach a helicopter under power from the front within view of the pilot, and must never enter the rotor disc without the pilot's positive permission; there is no negative signal. Having gained permission from the pilot, the tactical commander approaches the port side of the helicopter and every door and hatch is visually or physically checked. The tactical commander then enters the helicopter via the starboard sliding door, and plugs into the intercom to inform the pilot that all checks are complete.

Crew training and experience

Pilots

It is the responsibility of the operator (NPAS) to ensure that all pilots are competent and proficient to fly a helicopter and to deal with a range of in-flight activities and emergencies. This is known as an OPC and is usually accompanied by an IFP check. Additional training for the policing role is known as line training, and it involves instructing the pilot in the finer details of the policing role. Included are the positioning and manoeuvring of the helicopter for searches, pursuits, public order events and the whole range of other policing roles, along with procedures for opening the cabin door in flight for aerial photography. Training competence is assessed in annual line checks, which involve the whole crew flying with an instructor TFO or pilot.

Tactical flight officers

Training for TFOs is slightly different, because while there is a general assumption of policing knowledge and experience it is expected that new TFOs will have no prior aviation experience. Core training covers the majority of the aviation-related knowledge that is needed for safe operations in and around helicopters. Local procedures training looks at the specific operating environment within which the TFO will work. It covers the safety

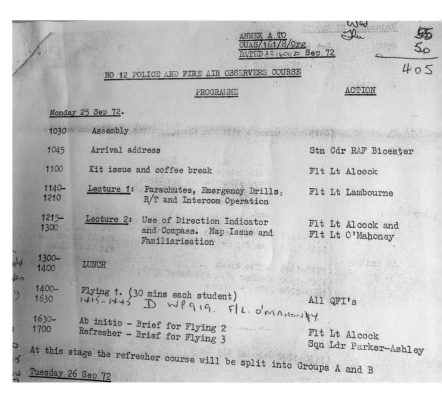

ABOVE An extract from the Air Observers' training course, 1972. The five-day course was aimed at police and fire officers.

BELOW An extract from the Air Observers' training course, 1986. This is more advanced, and covers aspects of safety, operations and tactics.

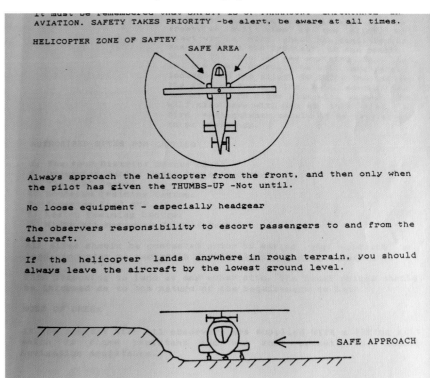

Putting skills into practice in a Safe environment.
Making & learning from mistakes

Exercise 1	Exercise 2	Exercise 3	Exercise 4	Exercise 5	Exercise 6
					Pursuits Practical
				Searching Practical	Searching Practical
			Moving Targets (vehicles)	Moving Targets (vehicles)	Moving Targets (vehicles)
		Moving Targets (people)	Moving Targets (people)	Moving Targets (people)	Moving Targets (people)
	Basic Search Techniques	Basic Search Techniques	Basic Search Techniques	Basic Search Techniques	Basic Search Techniques
Basic Camera Operation	Basic Camera Operation	Basic Camera Operation	Basic Camera Operation	Basic Camera Operation	Basic Camera Operation

features on a specific helicopter and the operating base, along with the operation of the role equipment that is fitted. This course qualifies the TFO to undertake the front seat TFO role and gives the knowledge and skills needed to successfully operate the camera system, navigate to task and support the pilot and crew from that front left seat. The aim in TFO selection is to identify those candidates who are most likely to complete training successfully and achieve competence within the amount of flying time available, typically 12 to 15 hours per student.

BELOW ASU crews undergo HUET. In this picture crew members are seen learning to use the short-term air supply system.

Tactical commander course

After about six months of operational flying, new TFOs are ready for the tactical commander course. This is again a mixture of classroom-based ground school and flying training, and involves around eight to ten hours of intensive flight training. Its aim is to ensure that the TFO is competent in the delivery of the full range of air support tactics. Training sorties involve a mix of other crew members and specialist resources who play out a full range of complex operational scenarios. It is during this training that the pressure is significantly ramped up, with TFOs operating at and sometimes beyond capacity at all times when they are flying. The instructors induce mistakes, force decision-making and create an environment in which the student tactical commander is able to make mistakes and learn from them in a safe and secure environment.

Other training, assessments, skills and qualifications that are needed by crews include aviation medical, fire training, emergency and survival equipment and procedures training, helicopter underwater escape training (HUET), specialist roles training and CRM.

Maintaining safe operations

The maintenance of safety at all times is crucial to the delivery of a consistent air support service. The history of police aviation in the

UK thankfully includes very few accidents, but from each of those that has occurred lessons have been learned – with the aim to be always as safe as possible. The business of policing can be risky, and every police officer enters the role knowing that it is impossible to guarantee their safety absolutely. There are, however, many measures that can be taken to enhance safety and reduce risk to as low as reasonably practicable in any circumstances. The world of aviation is different in many ways, as any accident or incident is one too many; but if a safety culture is maintained throughout then flying remains one of the safest forms of transportation. Police flying is dynamic and challenging, though: one minute you can be flying over crowded urban areas in among cranes and tower blocks, and then in the next you can be operating in large rural areas with uncontrolled airspace, changing weather and many other challenges.

So how do you keep air support operations safe and efficient at all times? The culture of any organisation is what runs through it like the marbling of a piece of rock: it touches everything the organisation stands for, what it does, how its people behave and above all how it operates every day without needing to think about it. It has been said many times that 'Safety is our number one priority', but those are just words: what does safety culture actually mean in practice? Police units, along with the other aviation services, operate safety management systems (SMS) that are mandated by the UK regulator, the CAA, and this means NPAS has to constantly re-examine its safety systems and processes to ensure they are fit for purpose.

It is essential that there is no difference between the standards of safety expected of a law enforcement aviation operation and those of a commercial operator. The SMS is underpinned by a positive and tangible safety culture that runs throughout any ASU, and it is the personal responsibility of everyone involved in that operation to make a positive contribution to safety. Police crews spend a lot of time learning about safety: they undertake CRM training and learn about the importance of communication and cooperation; how to work together effectively at all times. They examine aviation accidents, especially those

involving police aircraft, exploring the what, where, when, how and why of every example and looking at how they can prevent similar recurrences. Police officers are generally good at assessing risk: the National Decision Making Model is used by officers every day to support safe and effective decision-making in policing. The risks presented by the aviation

ABOVE ASU crews undergoing HUET, also known as 'dunker training'. In this picture they are being trained to use life rafts for survival in the event of ditching. *(Hugh Dalton)*

BELOW ASU crews undergoing fire refresher training. An initial course is delivered upon joining the ASU, with a refresher session every three years. *(Hugh Dalton)*

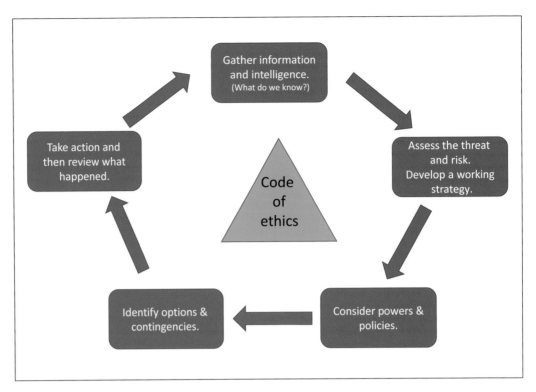

environment are different, and while a vast array of regulations and procedures exist to maintain safe operations, the knowledge of the humans who are implementing that safety system is crucial. Simple things that a career aviator would perhaps take for granted may not be obvious to a police officer who is entering the aviation arena for the first time. A good example of this would be the simple rules about who is in charge of an aircraft and the communication required before doing anything that changes the aircraft's status. If a pilot has signed for the helicopter and it is prepared for operations and on line, that pilot and the duty crew need to know that when they run out to the helicopter to take off they will find it exactly as they left it. A hierarchy of communication and permission is essential to ensure that this happens at all times, and this may not be immediately obvious to a non-aviator.

Police aviation carries some inherent risks: the delivery of a spontaneous emergency service at all times of the day and night, in congested airspace and often poor weather means it is difficult to plan in any detail before you start the task. It is very common that the crew don't know exactly where they are going, what they are going to do and how they are going to do it when they strap themselves into a police helicopter and go flying. Gaining confidence in safety is not easy, and to arrive safely on scene at a police operation, deliver a professional and effective service to colleagues on the ground and return safely to base requires a safety focus and team effort from everyone involved.

The Eurocopter EC145, in common with most police helicopters, is fitted with safety features such as TCAS, and when operating in controlled airspace the support offered by ATC to maintain airspace awareness is crucial. This is not always available, however, and police pilots and crews need to work well as a team. Police pilots are professional, skilled and experienced instrument rated pilots who have thousands of hours of flying experience and are trained to the highest standards to fly day or night. UK police aviation has evolved massively since the early trials using hired helicopters; the CAA has always specified that police helicopters are twin-engine and flown by experienced professional pilots with a minimum level of experience. The police helicopter crew of three, one pilot and two police officers trained as TFOs, are professional aviators; experiments with part-time or temporary TFOs in the early days exposed the gaps that a lack of knowledge, training or aviation experience brought. Professionalising the role of TFO was considered critical to the safe and efficient delivery of police aviation services. These days police helicopters are packed with an ever-increasing array of sophisticated technical equipment that when used effectively enables the crew of a police helicopter to deliver fantastic results. However, professionals are needed to do this.

Operating together ensures that there are three pairs of eyes looking out of the cockpit at all times. The pilot can focus on the safe flight of the helicopter while the two TFOs can focus on the police or other task in hand. Every task is tackled as a crew of three, and if any member of that crew is not performing well the safe and effective completion of the task is jeopardised. The workload in the cockpit during an operational task can be significant, and it is essential that each member of the crew is not only trained and competent in their specific role but is operating in accordance with standard operating procedures (SOPs). Many years have been spent working with crews to arrive at a standard method of operating each item of equipment and delivery of each tactical option. SOPs are exactly that; but they are not the only way of working. However, for most tasks everyone in the cockpit knows their role and what is expected of them if the SOP is followed. It is essential, for example, that the pilot positions the helicopter to enable the front seat TFO, who is operating the on-board camera sensors, to get the best view possible. Communication in the cockpit is also crucial, and throughout every mission the crew are constantly talking to each other as well as listening to the ATC and police radios. When training as a TFO the thing that most find hardest to master, apart from navigating in the dark from a map book on their knee of course, is how to listen to multiple radios and decipher the pertinent and important information from the routine and irrelevant. It is inconceivable to think that an untrained police officer would ever be put into a helicopter and expected to undertake the TFO role.

So what has this got to do with safety? If we accept that the effective operation of a police helicopter is whole crew responsibility, a TFO has a key role that maximises safety and creates this much-desired safety culture. Every

shift starts with a full crew briefing, during which the pilot leads the TFOs and the air operations officer (a dedicated ASU dispatcher) through the weather, NOTAMs, actions in the event of an emergency and aircraft status brief (i.e. flying hours available, maintenance status and any equipment defects or issues). This ensures that every member of the crew is on the same page at the start of the shift. Even if a crew is flying together for a number of consecutive shifts it is never assumed that nothing has changed since the last tour of duty. Complacency introduces risk and impacts upon safety; airspace awareness is essential to remain safe, and this involves every member of the crew. On one occasion a tactical commander spotted something out of the corner of his eye that turned out to be a homemade kite with 1,000m of thick twine hanging from it. It had blown away and become snagged in some railway power cables and was floating around, almost invisibly, at 800ft; it was waiting to be ingested by an engine or wrapped around a tail rotor. TFOs are not along for the ride: they have a significant role to play and are actively encouraged to keep a safety lookout and call air traffic and other hazards for the benefit of the pilot. Sure, on most occasions a pilot will have seen the hazard if a TFO has seen it, but a simple acknowledgement of 'Got it!' from the pilot gives the crew a warm and fuzzy feeling. One of the first things taught to new TFOs is that there is no such thing as a stupid call, and that they will never be criticised for making even the most obvious safety comment.

This briefing at the start of each shift or crew change is the foundation of the safety management system. There is also an emergency of the day, which is the topic of a pilot-led discussion either during the briefing or at some point during the shift. On the wall in the briefing room is a list of around 30 topics; on the 15th day of the month the crew discuss the 15th topic, for example. The list includes topics such as engine failure, inadvertent flight into instrument meteorological conditions (IMC), or unplanned flight into cloud as it may be better described, fire, generator failure, fuel management and so on, and the thinking is that throughout the course of a year crews will discuss everything on the list. Doing this

RIGHT The emergency
of the day is a list
of potential in-flight
emergencies that are
discussed during the
month by the crews
when briefing.

METROPOLITAN POLICE SERVICE
AIR SUPPORT UNIT

EMERGENCY
of the DAY

Day	Daily Emergency
1	Double Engine
2	Single Engine
3	Governor
4	Fire
5	Smoke
6	Tail Rotor
7	CRM Topic
8	Fuel System
9	Generator/Invertor
10	Hydraulics
11	Autopilot
12	BATT T°
13	Main Gearbox ChipOil /Press/Temp
14	Doors
15	Eng Chip/Oil Press/Temp
16	CRM Topic
17	Tail Gearbox
18	CAD Cautions
19	Lighting
20	Air Data
21	Antenna
22	Inadv IMC
23	Incapacitation
24	Ditching
25	Disorientation
26	FCDS
27	FOD
28	Icing
29	Comms
30	Vortex ring
31	Pitot/Static System

BELOW A
computerised version
of the emergency of
the day, as used at
NPAS London.

NPAS LONDON BASE
Daily Emergency Training
EC135 T2+

RANDOM **SPECIFIC**

TO INITIATE A RANDOM EMERGENCY

LEFT CLICK ON EITHER THE
RED OR YELLOW BUTTON ABOVE

(Red = CWP   Yellow = CAD)

TO INITIATE A SPECIFIC EMERGENCY

RIGHT CLICK ON THE GREY BUTTON
ABOVE. SELECT 'GO TO SLIDE'
AND CHOOSE YOUR EMERGENCY

(93 Slides)

sharpens the mind, educates the TFOs and
creates a feeling of joint ownership of flight
safety issues: the whole crew feel trained
and equipped to deal with whatever the flight
throws at them. This process of discussing
emergencies regularly and routinely is practised
in a number of ways. For example, a computer-
based system developed at one of the NPAS
bases offers the crew an opportunity to select
random emergency and flight safety topics to
discuss at the briefing each day.

Assessment of risk is also an essential
component of any safety system; formal risk
assessment processes are undertaken to try
and address every one. Trained personnel
try to identify the major risks and understand
them before implementing control measures to
manage and reduce them. There is nothing in
the risk assessment process that states that
if there is a risk that can't be eliminated then
you can't accept the task; it is simply better
for the crew to know and understand that
the risk exists in the first place. The heart of
any SMS is the involvement of all personnel
in the identification and management of risk.
The seven most significant risks at any air
support base are identified and referred to
as the Significant Seven. Posters are then
produced and placed around the base so that
staff are constantly reminded of these risks and
consider their implications. A good place to put
any notice is on the back of a toilet door: this
appears to be a place where staff have time to
read notices more thoroughly than when they
are on other noticeboards. At NPAS London the
significant seven risks are identified as:

- **Distraction** – extended period of change,
  job security, lack of communication or
  rumour.
- **Collision with elevated structures** –
  cranes, buildings, pylons and wires.
- **Airborne conflict** – on departure, in the
  vicinity of heli-routes, outside controlled
  airspace.
- **Special roles** – lack of power, obstacles,
  procedures.
- **Helipad hazards** – unauthorised access,
  noise, downwash damage.
- **Unmanned aerial vehicles (UAVs)
  (drones, etc.)** – small and hard to see,

increasing in numbers, operating at increasing heights, not all NOTAMed.

■ **Maintenance hazards** – climbing on aircraft, hangar obstacles, fluid spillages.

As police aviation can be unpredictable and hazardous, there are some tasks and circumstances that are unusual, unforeseen and cannot be reasonably planned for or briefed. A process of in-flight dynamic risk assessment seeks to address this. The crew completes a dynamic risk assessment matrix prior to undertaking any non-standard or potentially hazardous task: this could be something like an ad-hoc landing or an out of area task. This forces the crew to methodically consider and identify areas of highest risk and determine what they can do, at that time and place, to manage those risks. Each risk identified by the crew requires consideration of control measures to acknowledge it and reduce it to as low as reasonably practicable. There is nothing in the process to say that they cannot undertake a high-risk task, but if they do so they will have identified and managed the major risks. The crew record what they have considered and the control measures they have implemented on this dynamic risk assessment.

ABOVE **The Significant Seven risks poster, which is displayed throughout the ASU base to remind staff to be vigilant and retain awareness of risk.**

LEFT **One of the safety notice boards that are located at every NPAS base.**

Chapter Five

Tactics

Air support delivers a range of policing tactics that reduce public and policing risk, enhance capability, add value and support local officers in the fight against crime. The vast majority of air support work involves searching in one form or another, and the crews use a mix of policing skill and technology to maximise operational success.

OPPOSITE The Met's EC145 helicopters took to the skies over London for around 2,700 flying hours each year and attended around 7,500 operational tasks. Here G-MPSA is pictured above central London.

What does a police helicopter do?

The roles undertaken by police helicopters have evolved since using aviation to support policing was first considered. Primarily introduced as a public order resource rather than to help with crime fighting, helicopters operated as a form of flying club during the week in the early years of air support, as long as they were available for demonstrations and events at the weekends. Police officers' ingenuity and creativity meant that helicopters were quickly put to other uses, searching for criminals, engaging in pursuits and coordinating raids. Early crews used binoculars (which were sometimes gyro-stabilised) and hand-held cameras, and did so very effectively given the obvious challenges. The arrival of stabilised multi-sensor camera systems, which could be operated day or night, opened up new opportunities for police aviation, and around the country ASUs sought to add as much value to their local force as they could through the use of helicopters and their role equipment.

Air support roles such as searching, managing vehicle pursuits, creating aerial imagery, managing critical and firearms incidents, and providing security escorts, surveillance and cover at public order events all evolved as crews experimented with and promoted the use of air support to bring something extra to these activities. The flying club was no more, and ASUs fast developed a range of tactical options that forces increasingly relied upon and missed when they were not available. The value of air support has been the subject of debate over many years, with a number of TFOs and pilots attempting to measure and articulate the value that having air support brings to policing in terms of reduced risk, increased efficiency and cost effectiveness. In 2000 Simon Mitchell, a pilot with SERPASU, compiled one of the only formal pieces of work on the subject, which was a thesis for his university degree. He looked at the cost effectiveness of air support by applying business-related cost benefit analysis techniques. His work presented the view that air support added value in three ways: saving time, improving capability or effectiveness, and reducing overall risk and liability.

The actual benefit derived from air support is difficult to measure, as it allows things to be done that would not normally be possible: ground-based resources and capabilities simply don't exist in the quantity required. On the other hand, air support can sometimes deliver more than is strictly necessary, as larger areas are searched because doing so is easy. There is, however, no doubt that value is added: officers on the ground gain huge confidence and benefit from knowing that the ASU, the 'eye in the sky', is looking down on them, ensuring they remain safe and maximising their chance of success.

NPAS looks to assess the viability of any given tasking request using a model that considers the threat, risk, harm and vulnerability present in given circumstances, measured against the chances of success of air support deployment. For example, searching a crowded street during daylight for a male with limited description or a white van with no distinctive markings, registration number or direction of travel is like searching for a needle in a haystack; such tasks are not likely to be viable. Searching for a suspect seen to run into rear gardens after a burglary and contained by police on scene has much more likelihood of a successful outcome. NPAS use this methodology as they aim to filter the many thousands of requests for air support service that they receive throughout the year, and ensure that the resources available are deployed to those events that present the highest risk and/or where the outcome has the greatest chance of success. The removal of crown immunity from the police in terms of health and safety at work legislation brought about significant changes to the police risk appetite. It was no longer acceptable for a police officer to climb on to a rooftop, enter water or go on to railway lines without a complete risk assessment taking place. The use of air support to mitigate these risks is clear: a rooftop can be cleared in seconds, a railway line searched quickly over many miles, water searched and rescues coordinated, without the need to put ground officers at risk or spend hours deploying officers with appropriate training and equipment.

As air support has evolved, the range of roles undertaken, the tactical options, has been relatively consistent as ASUs in the past and now NPAS seek to maximise the benefits of each and every deployment. There is no doubt that the development of UAVs and their growing use by police has changed the appetite for air support and will continue to do so in future. Local policing issues, small area searches, aerial imagery and other pre-planned activities have all benefited from air support by drone; however, the role of police helicopters and fixed-wing aircraft in dynamic and high-risk tasks remains consistent, and demand today continues to be high. The crews of police helicopters have always pushed the boundaries and attempted to look at new and inventive ways in which they can add value to policing, some being successful and others not.

In order to understand what a police helicopter does at any given incident, we can examine three of the most common tactical options, going inside the cockpit and looking at what the crew are doing. This shows what air support can achieve when tasked with a search, vehicle pursuit or public order incident.

Searching

The use of a police helicopter for searching accounts for about 60% of all air support tasking. Over many years in the Metropolitan Police ASU searching was split between the following different types:

- Criminal suspect searches.
- Missing, vulnerable or injured persons searches.
- Security searches.
- Vehicle searches.

While the proportions of search types may vary from region to region, the quantity of tasking devoted to searching is largely unchanged. Each has a different potential outcome, and as such the method of search and purpose of search will differ depending upon what the crew is looking to find. A criminal suspect search is about finding a person who is known or strongly believed to be present within an area and is hiding to evade detection. A missing person search is about scanning large areas for a person who may or may not have intentionally gone missing: they could

be stranded, lost, disorientated, confused or otherwise in real danger, and a systematic search from the air aims to find them. Security searches are precise and methodical, as large areas, frequently rooftops, enclosed or otherwise inaccessible areas, are searched and cleared or items are brought to the attention of specialist search teams. Vehicle searches are rare, but on occasion the nature of criminality or distinctive description of a vehicle makes an airborne search a viable option. From the air crews can see into enclosed and inaccessible places in a way that might prove impossible from the ground.

As there are so many variables in relation to searches, crews need to obtain as much background information as possible before determining how best to approach one. A garden search, with sheds, garden furniture and canyons between buildings to contend with, requires a specific technique to ensure that every angle is covered and the area is searched thoroughly. An open area search, on the other hand, may well have fewer physical obstacles to overcome but must nevertheless be methodical and organised to ensure that nowhere is missed. There will always be places that cannot be seen from the air, such as under hedgerows or trees in summer, and these areas must be identified so that they can be searched by police dogs. If 90% of an area can be searched from the air, then the ground resources can focus upon the remaining 10%: this increases the chances of success.

A search of rooftops, railways and water presents different challenges, and as such a variety of search methods and techniques are used. The fact is that the crew can plan a search in advance but it is only upon arrival on scene that they can truly examine the task in hand and plan and brief for its completion. The most important thing in the cockpit is communication between pilot and crew to ensure everyone is at the same place and understands exactly what they were doing and how they are going to do it. If the search task is urgent and requires a response as soon as practicable, the time for planning pre-take-off may be limited and all planning is done en route and upon arrival on scene. Examples of urgent searches are those for injured or suicidal persons where life is at risk or searches for criminal suspects where significant ground resources are awaiting air support. Searches such as security searches and missing or vulnerable persons searches are more pre-planned and a lot more time can be spent pre-take-off determining where the crew are going, what they are required to search and how they plan to search it.

Enclosed area search

One of the most common searches is for a criminal suspect who has either been chased by police officers and lost or disturbed following a burglary or another criminal act. Such foot chases frequently end up with suspects entering enclosed rear gardens and seeking to hide from police amid the mix of sheds, furniture, trees, shrubs and other things, such as children's toys and trampolines, which are in many suburban back gardens. Typically, information will be given that the suspect was seen to enter rear gardens and was last seen at a specific address, 55 Acacia Avenue for example. Using the mapping system on the helicopter it is possible to zoom into a building

outline layer and identify exactly where that specific address is, and then to determine the scale and extent of the search area. Most commonly a specific street will back on to at least one other street, and as such it is possible to define an initial search area as being the rear gardens enclosed by those streets. The area may not seem that big at first glance, but rear gardens present a number of challenges to the helicopter crew: they are complex and typically comprise a number of individually fenced gardens, each of which contains a multitude of items that could provide cover.

When engaged on a garden search the pilot will most commonly put the helicopter into a right-hand orbit. Most air support work is done using a variety of different right-hand orbit patterns, which aim to present the pilot with the best field of view for identifying and holding points of reference and to put the camera system, mounted on the front of the right-hand step in the best position to obtain a good view of the search area. The aim is to position the camera so that it is looking roughly 90° to the right in azimuth and with an elevation of -45° degrees down. Clearly this position is not held, but it provides a starting point for the camera

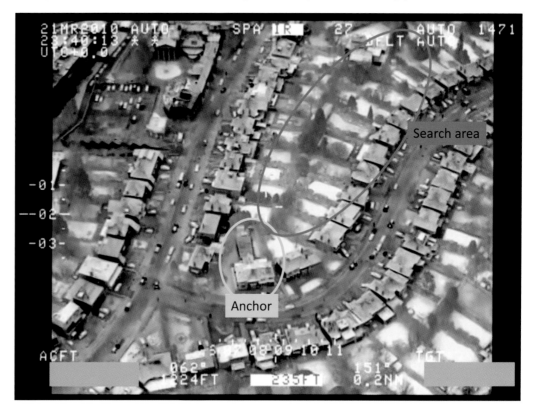

RIGHT Defining an anchor and the search area in a search scenario.

operator who is sitting in the front left seat and as such is largely unable to see the target. By day or night the camera operator and crew will agree a point of reference as a starting position for the search and will then work methodically from there, looking at each and every garden in detail as they move around the search area. The point of reference, known as an anchor, may be a specifically shaped house, a distinctive item of garden furniture or another feature that can be easily identified visually in daylight or on IR imagery at night. By working from the anchor point the camera operator will be able to ensure that every inch of every garden has been covered.

The benefits of orbiting a search location are numerous. First, it is more fuel efficient; secondly, it enables the camera to see every aspect of a search area; and thirdly it spreads around the noise so as to be not quite so intrusive. Crew members are always very aware that most of these searches happen at night, and the noise that the helicopter makes from 1,500ft over rear gardens is significant, and likely to wake up every occupant of the houses around the search area. For this reason, the crew work quickly and efficiently to ensure that all the initial search area is covered, and that progress is communicated to officers on the ground. On the initial couple of orbits of the search area the crew will be looking at a wide-angle overview. In daylight a mix of daylight and IR camera is used, along with the crew's eyes out of the window. It is amazing what the 'Mark 1 eyeball' can see that a sophisticated camera sensor can't! They are looking for anything obvious: who is in the rear gardens, what heat sources show up on IR and can they resolve the search quickly and efficiently. Remarkably this sometimes actually happens, and the suspect is either located or makes a run for it as soon as the helicopter arrives on scene.

More commonly, however, the crew will then commence a more detailed search, focusing on each garden in turn, moving between wide angle and narrow angle on the camera sensor to investigate and explore every area: underneath and behind garden furniture and in hedgerows and undergrowth hidden from view of everything except the IR sensor. With

ABOVE The area between houses can be hidden from view in a standard orbit; these hidden areas are referred to as canyons and can only be viewed from a specific aspect. The pilot places the helicopter so that the camera sensor can look directly into the canyon to clear it. This is typically a slow forwards creep with the houses at 90° to the right.

BELOW The crew is looking for anything in the image that does not look right. In this picture a sliver of heat can be seen coming from the door of a garden shed, which turned out to be the suspect inside. Crew members need to decide if it is worth sending a dog handler or ground officer to check out any potential heat source located.

RIGHT This suspect thinks he is invisible as he lies on a flat roof in the dark, hidden from view from the ground.

BELOW Searches of complex environments can prove a real challenge. In this case the suspects had broken into a goods yard containing a number of HGVs. The crew spotted a heat source hiding and moving under the front of a lorry. They began the process of talking in ground officers while maintaining an observation on the suspect's likely hiding place.

ABOVE The suspect broke cover as he heard the police approach, and in complete darkness thought he could move to hide under a different lorry. He was clearly visible using the IR sensor.

LEFT This suspect was chased after committing an aggravated burglary armed with a weapon. The back gardens search revealed a heat source in a shed, the door of which appeared to be warming up as the crew looked at it. Armed officers and police dogs were talked in to search the shed and make an arrest.

each orbit the camera operator will work up and down the gardens, moving from one to the next and making sure that every heat source detected is identified and eliminated. In the early days of IR the sensor was able to detect a heat source as a black or white blob in contrast to its surroundings, depending upon whether the camera operator was working black or white hot. Resolution and detail were not present, and it was very difficult to differentiate between a compost heap (which gives off a lot of heat), an animal such as a fox or a suspect. The crew had to decide if they would send in a police dog to check out the heat source or alternatively illuminate it with the searchlight and see what it was on the daylight camera. The current IR sensors are so powerful that they can see huge amounts of detail. Animals are easily identified and human heat sources more often than not stand out. It is possible to look at any given heat source and quickly and easily eliminate it as a pool, animal, garden ornament or something else other than the hiding suspect. In simple terms, crews are looking for anything within the scene that does not look right, something indicating hot that should not be doing so, whether there is a sliver of heat between two buildings that should not be there. Crews are trained to undertake this live imagery analysis in the aircraft, looking at each scene on screen and discussing what may or may not be worthy of further investigation. The use and interpretation of IR imagery will be covered later in this chapter.

Open area search

Searching open areas requires different tactical considerations. A crew needs to decide if it is best to break the search area down into smaller more manageable sections and cover each one using the orbit or race track flight profile, or if they want to search the whole area as one, adopting a ladder or creeping line search pattern (a description of these flight patterns is included later). The principles are the same: the first sweep is a wide angle search of the area looking for anything obvious, something moving, something that stands out against the background on IR or something that just catches the eye of the crew.

ABOVE In this picture the wheelie bin on the right is black, which indicates it is very hot compared to the one on the left, which is grey and colder. This could be composting vegetation, but looking at all the bins in the street the striking heat of this one suggests a human suspect inside. Officers were talked in, and a suspect was found hiding inside.

BELOW The panel next to the window in this shed is showing an ever increasing black spot, which turned out to be a burglary suspect inside.

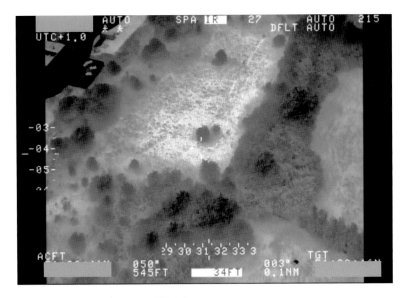

LEFT Here an open area covered by trees and bushes is the search area for a vulnerable missing person. Police had clear information that she was lost somewhere in this area, in complete darkness. The IR camera is capable of showing the smallest heat sources, and the crew spent many minutes looking at rabbits and foxes. The search area has to be divided into segments to ensure it is all searched.

CENTRE Eventually a heat source was spotted under a tree. This was examined from all aspects and seen to be a person lying down and waving. This was the missing person, who had somehow reached the middle of this inaccessible area.

BOTTOM The crew decided to undertake a night-time ad-hoc landing and to get out and rescue the female themselves. First aid was given and she was airlifted to a place of safety, where she was placed in the hands of the ambulance service.

Having satisfied themselves that there is nothing obvious in the search area, crew members can start their planned, methodical and systematic search of the entire area.

It is clear that a police helicopter provides a valuable search capability that is difficult to rival on the ground. Helicopter crews can search huge areas very quickly, clearing the vast majority of them with ease and focusing ground resources, such as police dogs, on to areas that cannot be easily searched. There have been various research projects completed over the last three decades looking at the effectiveness of police aerial searching compared with conventional ground search teams. There is no doubt that a competent and effective ground search team managed by a police search adviser (PolSA) can effectively search a given area on the ground with a great degree of success. The use of police search dogs can speed up the process and they can cover large areas, including potential human tracking, but the use of a helicopter to complete the search has in all studies proved to be the most efficient and effective method. In order to determine the value of helicopter use, it is important to look at the relative costs.

How many police officers deployed for how long would be needed to search a large area on foot? How confident can you be that a police dog and handler has searched an entire area, and how many dogs and handlers would the search take? The deployment of a helicopter, which can search a large area very quickly, has been found to be cheaper by a factor of up to ten when compared with the cost of deploying ground resources in the numbers required. It is for this reason that a combination of air and ground resources is preferred for such searches. If a helicopter can clear 90% of a given area with ease, then the focus of ground resources can be on the remaining 10%. Such areas can include thick undergrowth, trees in summer with heavy leaf cover and areas that are beneath bridges or otherwise hidden from view.

Search Areas

5 km radius = 78.5 square km

½ km radius = 0.78 square km

Size of search area
It is important to decide on a realistic search area as starting too far from a target location creates a potentially massive area to search.

ABOVE AND BELOW A series of diagrams that explain search patterns and areas.

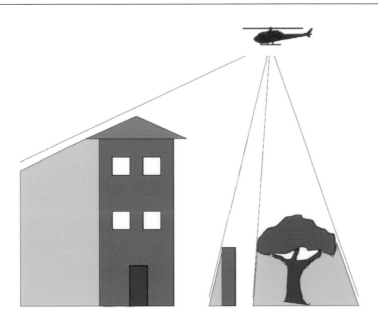

Search every aspect
If the crew only look from one aspect, the buildings, trees and obstacles on the ground will shield certain areas from view. Operating in the orbit helps the camera operator to search every available aspect clearing the whole search area.

Moving from an orbit to a slow creep ahead it is possible to angle the camera so that it can see clearly into every alleyway in a way not visible from the orbit. Sometimes it is also necessary to look straight down into the canyon space.

Search canyons and alleyways
The proximity of buildings to each other will always generate alleyways and spaces between buildings that can only be seen from one specific aspect (the red areas).

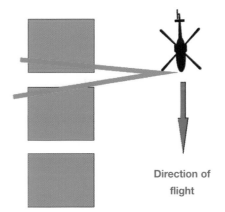

Direction of flight

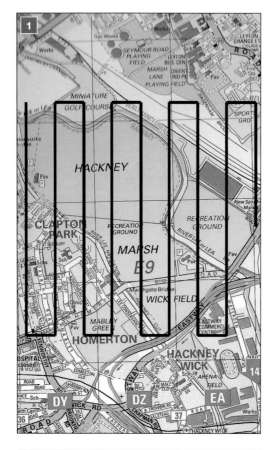

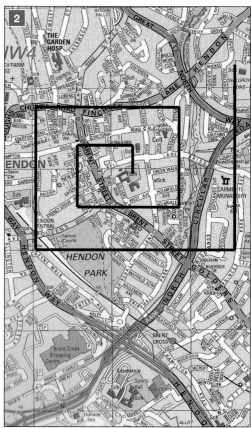

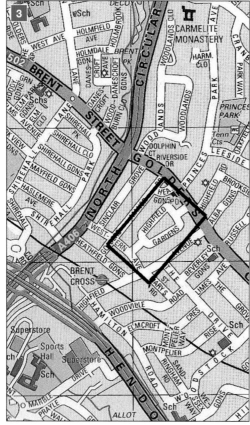

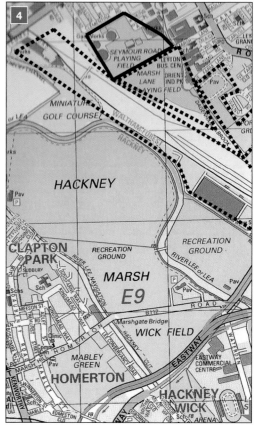

There are a number of generic search patterns that are contained in the air support tactics manual. However, the fact that every search is unique means that they are seldom used in isolation and are commonly blended.

1 Rising ladder or creeping line

This pattern sees the helicopter fly a specific route backwards and forwards over a search area. It is used to search a large open area with confidence, and ensures that every part is covered by the camera.

2 Expanding square, spiral or datum

This pattern is used when a starting point is identified and the search is to be expanded from that point. This could be a school or nursing home from which a child or elderly person is missing and where they were last seen, or it could be the home address of a missing person.

3 Race track or orbit

This is used to search an enclosed area or gardens. It ensures that the crew is able to search the entire area, seeing between buildings and into enclosed canyons.

4 Sector

This is used to search large areas, which are then broken down into smaller more manageable areas. Each is searched and cleared in turn.

5 Spiral

This was developed to cover a crime scene and its surroundings in the aftermath of a crime or to capture activity within a given area. The crew typically searches and captures the actual scene of the crime first, then moves away in a slow spiral search pattern before returning to the crime scene. Everything that is present within the given area is captured.

The technology fitted to modern police helicopters enables the crews to shade on a map the areas that the camera has been pointed at. There is no guarantee that anyone was looking at the image when it hit a specific area, but close crew cooperation means that crew members can record with confidence the areas that have been searched.

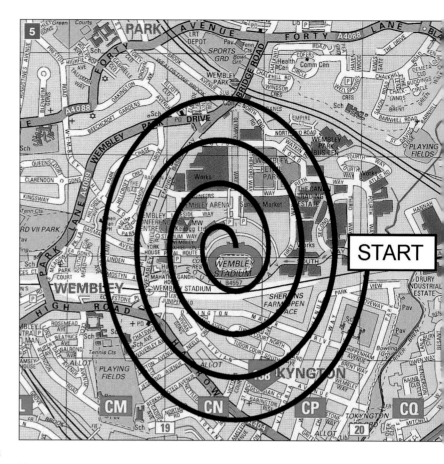

High-risk area Searching from the air is by far the biggest part of police air support tasking. There are a multitude of ways in which a helicopter crew can deploy the sensors that are carried to search and clear any given area. The aerial search reduces risk and supports ground officers so that they can quickly and efficiently return to other duties. High-risk areas such as rooftops or other high objects such as tower blocks, cranes and railway lines can be searched with ease. If air support is not available, it is assumed that the police might choose not to search and clear these areas, or to do so with the deployment of specially trained and equipped ground-based officers.

Railways – The report of suspects on railway lines always causes huge problems for the British Transport Police and for railway operators: the risk of injury is high, as is the cost of disruption and inconvenience to rail users. The ability to quickly and effectively fly along large stretches of railway lines searching both sides means that the risk can be massively

LEFT Industrial premises and abandoned buildings present significant risks. Their rooftops are often used as a means of access, and height is a real challenge for ground officers and health and safety. Searching from the air is quick and effective. In this case suspects had been seen on CCTV entering disused factory premises and were believed to be on the roof.

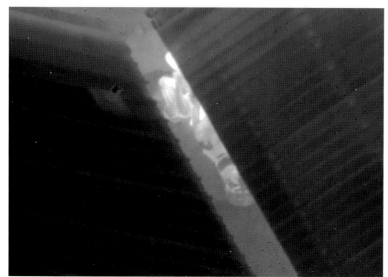

CENTRE They were soon located lying in a gap between two sloped roofs. The next challenge for police was their safe removal. The Nightsun searchlight was used to illuminate them and indicate that they had been seen; this prompted them to come down.

BOTTOM Some suspects think that if they can't be seen from the ground they are safe. This is not the case, as this picture clearly shows.

reduced, and the need to stop trains or place them on caution diminishes.

Water – The ability to search water has always presented a challenge for police officers on the ground. The report of a person in the water is a common one in some areas of the country and the resources available to police in these circumstances are limited. The Royal National Lifeboat Institution (RNLI) and the Maritime and Coastguard Agency offer a search and rescue capability, but in an inner city or urban environment where there are canals, ponds, lakes and inland waterways the benefits of deploying a police helicopter are great. Using mapping technology it is possible to highlight an area of probability within which a search can be focused. The TFO can enter the starting point (the last position the person was seen in, which could be on a river bridge, for example), the time elapsed since they were last seen and the direction and speed of tidal flow. This generates an expanding set of concentric circles that show how far the person might have travelled, and this can be used to great effect by the crew – who quickly search the point of entry to the water, then move away to the edge of the most distant probability circle, then search back to the point of entry. The coordination of this search with other rescue authorities, such as the RNLI, is crucial to maximising the chances of success.

Overall the use of air support for searches is by far the most common use of police helicopters, and the results obtained have always withstood scrutiny and analysis. If an area can be cleared with a high degree of confidence by the crew, then ground resources can quickly stand down and go about their other pressing duties. This is an efficient and effective use of resources and may be measured as the successful deployment of air support. If the crew finds a missing person or suspect and they are rescued or brought to justice this is an obvious measure of success, and one that is regularly promoted by NPAS via social media.

Vehicle pursuits

Vehicle pursuit management is perhaps the most exciting and dynamic task that a police helicopter crew can undertake. The term vehicle pursuit is used to describe the situation after the driver of a vehicle fails to pull over and stop when legally requested to do so by a police officer driving a police vehicle. The reasons behind failure to stop are varied, but the result can be extremely hazardous: drivers take risks, drive at high speed and ignore road signs and conventions in their effort to escape. The management of pursuits is therefore one of the most risky roles that police drivers undertake: speeds can be high, the actions of the driver being pursued are unpredictable and the risk to innocent bystanders is increased. All police drivers are highly trained, but only those trained to an advanced level and authorised as pursuit drivers are permitted to pursue. Having said that, any police driver may be put in a position where they have to try to stop a vehicle, and not all will be pursuit trained and authorised. This presents policing with some challenges. The pursuit is therefore managed in phases, with the initial phase being from the point at which the vehicle fails to stop until tactical options and trained drivers are in place to bring the pursuit to a safe and satisfactory close; at that point the pursuit is officially

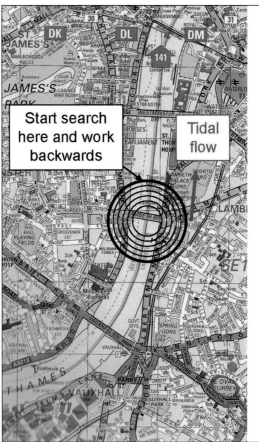

Start search here and work backwards

Tidal flow

Area of probability search

When trying to determine how far a person may have travelled since their last sighting, the mapping system can draw an area of probability on to the map. By selecting a starting point, a speed of travel and a starting time, the map can draw an ever expanding area of probability.

When searching a large river for a person seen to fall or jump in, the tidal flow and speed can be used to determine where the crew should start their search and how far away from the bridge they should do so.

BELOW Searches of rivers and waterways can be a real challenge. In this image a man had fallen into a river. His head was noticed by an eagle-eyed TFO. The heat source in the centre of this picture is a head, and as the man flailed around to stay afloat the IR sensor detected the slight changes in the water temperature around him. He was safely rescued when the crew talked in the RNLI lifeboat that was also searching for him.

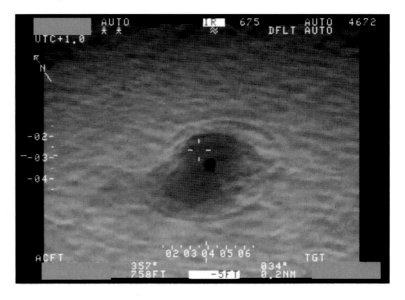

authorised. The assessment and management of risk to all parties involved is paramount, and unless the police have a tactical option to resolve the pursuit it will not be authorised and will be terminated immediately.

The use of air support is a key consideration in these situations, and its availability adds significantly to the chances of safe resolution. Across the country there are pursuits taking place every day, with a large proportion of them lasting for just a few minutes before the vehicle stops, crashes, is abandoned or is lost. As such, the chance of air support getting there and supporting pursuit management is slim; but all pursuits should be routed to air support for a response if practicable. The benefits of this are numerous: first, the helicopter crew can track the vehicle on camera, downlink the live imagery to the control room and record exactly what is happening. Secondly, because the TFOs can track the vehicle involved the police vehicles engaged can withdraw slightly, as they no longer need to retain direct sight of the suspect vehicle. This removes some of the pressure from the situation, as the suspect driver no longer has blue lights and sirens following closely and is more likely to either

slow down or decide to abandon the vehicle, or decamp as it is commonly known. The combination of resources used to manage the pursuit aims to reduce the speed, pressure and risks. The technology on police helicopters can be used by the TFOs to great effect, indicating how fast the vehicle is travelling and exactly where it is on the map at any time.

En route to the pursuit

Pursuit management is one of the few occasions where the helicopter can be tasked almost immediately by the requesting force using a radio hailing group; time is of the essence and the quicker the crew get into the helicopter and take off the better the likely outcome. The initial steer will most likely be a page reference number, road name, direction of travel or nearest junction, for example 'M1 southbound approaching junction 5'. The aim is to give the pilot and crew as much of an idea of the initial direction of travel as is possible to aid communication with air traffic and select the initial direction of flight. A pursuit situation is dynamic, and unlike other tasking where the location is largely fixed and known, the location of a pursuit is constantly changing.

RIGHT Pursuits of mopeds and motorcycles are exceptionally challenging for the police and the risks are high. In this case a suspect on a motorcycle is easily followed by the helicopter crew.

CENTRE Some pursuits develop to the point where the risk of continuing is too great and they are terminated. The helicopter is able to follow as the pursuing police vehicles withdraw. This often causes the driver to slow down and even abandon the vehicle and run off (decamp). This is not always the case, however, and sometimes the vehicle continues at speed, in this picture at night down a residential street.

BOTTOM The end result can be catastrophic, and in this image the suspect vehicle has clipped a parked car at speed, causing a significant crash. The IR picture shows the impact of the crash and debris as it falls from the vehicles.

After take-off the main priorities of the crew will vary depending upon their role in the helicopter, remembering at all times to keep each other and especially the pilot in the loop as they transit towards the pursuit location.

Immediately after take-off, the front seat TFO works to set up the camera system, making sure that it is on line, that the IR sensor is cooling correctly and the imagery is focused. The camera sensor is gyroscopically stabilised, and there are a significant number of electronic features that must power up and align to enable the camera to work in automatic mode. This means not only that the camera will be using geo-stabilisation to hold a specific position on the earth, making it much easier to operate, but also that it will interface with the mapping system effectively. Having sorted out the camera, the front seat TFO focuses on listening to radio updates and assisting the pilot in adjusting the heading. In some cases the last junction and direction of travel will be located on a map, and local knowledge will be used to identify the likely future position of the suspect by looking out of the window. At night this is slightly easier, as the blue lights of pursuing police vehicles show up better against the darkness. There is, however, a danger in simply chasing blue lights, as at any moment it will be possible to see blue lights of police cars, fire engines and ambulances that are travelling across the city and going about their unrelated daily business.

The tactical commander makes sure that the appropriate radio talk-group is selected, so that updates on the pursuit can be received, and also makes sure that the appropriate TAC radio is selected by the front seat TFO for recording by the video recorder. Most pursuits are conducted on a force-wide talk-group, meaning that specialist resources, including the helicopter, can communicate effectively via radio. Identification of the pursuit location is key and the mapping system can be used to assist with this. The tactical commander can use the joystick on the mapping display to move the cursor towards the target location that is being given by the pursuing police vehicle, and can adjust it as the pursuit progresses. Having selected the most recent target position, coordinates can be sent to the camera system using the SEND POS command and then the front seat TFO can slew the camera to the target coordinates sent to it by the map, using the SLEW button on the hand controller. This process can take place as frequently as required. Communication between the TFOs and the pilot is almost continuous at this time, as the pilot adjusts the heading to follow the pursuit and the TFOs work together to get themselves and the technology established, ready for arrival on scene.

The pilot listens intently to ATC radios and communicates the latest information and direction of travel throughout the transit. During most tasks the pilot is able to give ATC a geographical location and accept instructions regarding operation on task without interfering with other air traffic. The ATC clearance may be on the lines of 'cleared to transit direct to page 119, not above 1,500ft and not to operate further south than the A40'. Because of a pursuit's dynamic nature, it can stray easily into controlled airspace – with proximity to other air traffic being a real problem. Police pilots build a strong relationship with ATC, and this is built upon trust and confidence. The process of negotiating access to controlled airspace is two-

way and there is a real need for give and take on the part of police crews. Their helicopters fly using the flight category Bravo (CAT B), which allows them priority access to locations as an emergency service, but with reasonable holds and adjustments to work around other air traffic. It must be remembered that commercial air traffic is of real importance, and it is simply not acceptable for police helicopters to disrupt the airspace to the extent that commercial traffic is diverted or delayed – unless it is a life or death situation. In such a case, the police crew through the pilot can declare the flight to be category Alpha (CAT A): this means the helicopter will be allowed direct access to the airspace. Air ambulance helicopters transiting to an emergency or with a patient on board will use CAT A to quickly navigate airspace, but at other times they will use a lower category. This cooperation between the various users of complex airspace is essential, as delaying or diverting a single flight can cost thousands of pounds and inconvenience hundreds of people. The use of CAT A is a rare occurrence for police helicopters, though.

Activity on task during a pursuit

Once established on task, crew members need to pause and ensure they are ready to take over the management of the pursuit. There can be a tendency for officers on the ground to breathe a sigh of relief when the helicopter arrives and they hear that familiar call sign over the radio. There is no doubt that the chance of bringing the pursuit to a safe conclusion is enhanced by the presence of air support, and the risks are reduced for everyone involved. Before they can effectively take over the commentary on the pursuit, all three crew members need to be ready and fully aware of what is going on. The front seat TFO must have the pursuit on screen and be happy that the correct vehicle is being followed, the camera system is functioning correctly and they are ready to downlink live imagery of the pursuit. The pilot must have the pursuit visually identified outside the cockpit window. A mix of camera imagery and real world view ensures that the helicopter is positioned correctly at all times to keep the vehicle pursuit on camera, and that it doesn't disappear behind buildings or go outside the camera's field of view. If the pursuit gets on the wrong side of the helicopter or the helicopter is too close overhead, control of the camera system can be challenging or even impossible; it might even cease to function correctly and 'topple'. The pilot plays a key part in pursuit management, and ensuring they are ready is essential.

The tactical commander takes on all or part of the pursuit commentary to assist ground units in focusing on their driving rather than where they are. To be ready to take on the pursuit the tactical commander needs to identify the pursuit out of the window, select the correct radio talk-group on the TAC radio control head, ensure that the video from the camera and audio from the TAC radio is being recorded, ensure that the video imagery is being downlinked and that the mapping system is selected to display camera point of impact, actively working to follow the pursuit location. When everyone is ready, the tactical commander announces arrival on the radio and offers to take over the pursuit commentary. From this point onwards, provided every crew member does their job, the management of the pursuit is seamless – and ground resources can slow down and withdraw slightly.

It is difficult to describe what happens in the cockpit during a pursuit. Communication on the intercom between the crew members is relentless, as they all work together. This is balanced by clear and effective communication to ground resources and control room of pursuit location, estimated speed, direction of travel, identified hazards, offences committed or driver conduct, road conditions and likely future tactical options. A well-managed pursuit involving air support should appear calm and controlled to the control room operators and police drivers engaged in the pursuit. The police helicopter's presence and calm and assertive radio communications from the tactical commander are reassuring. The absence of the pressure that is present on the ground enables the crew to work together for maximum success.

Manoeuvring of the helicopter can be dynamic during a pursuit, with a series of aerial moves that resemble aerial ballet. The

ABOVE Other pursuits end up with bizarre and very dangerous behaviour. In this image the driver has stopped on the M25 and after a 360° turn has begun to drive the wrong way against the oncoming traffic. The pursuit was terminated, with the helicopter maintaining observation.

aim of the pilot is to keep the pursuit on the right-hand side of the helicopter and in front of it: the camera on most police helicopters is mounted either on the nose or on the front right step, so it works best looking down at around -45° elevation and out to the right in an arc between 40° and 110° in azimuth. The camera does function outside these areas, but it is harder to control and the risk of losing the target is greater. If the vehicle turns right there are no major issues, as the helicopter simply flies around the right turn in an extended arc, but left turns are dealt with differently. In order to maintain camera position, the helicopter undertakes a rapid 360° turn to the right, keeping the camera on the target vehicle and then repositioning to the left and behind the pursuit. This aerial manoeuvring can be more dramatic when the pursuit is in an inner city, as the buildings generate canyons that obstruct the view of the camera; as such, the pilot is constantly repositioning the helicopter and tightening the turns to ensure that a clear view is retained at all times. This rate of movement and the sheer amount of work taking place within the cockpit makes a vehicle pursuit one of the most highly pressured and dynamic tasks

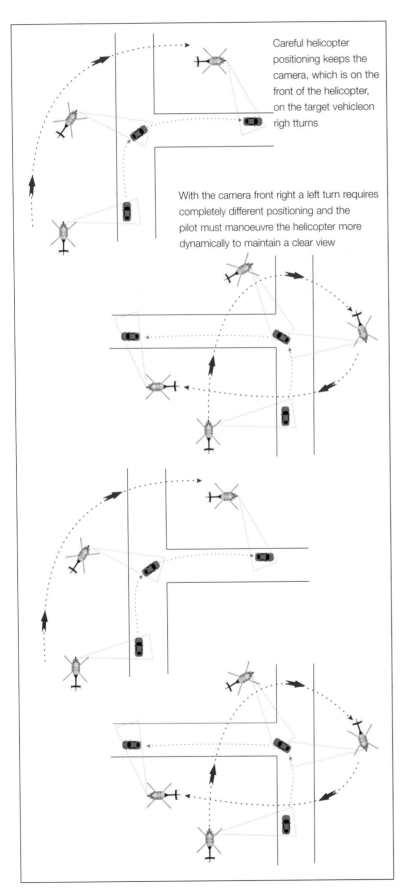

Careful helicopter positioning keeps the camera, which is on the front of the helicopter, on the target vehicle on righ tturns

With the camera front right a left turn requires completely different positioning and the pilot must manoeuvre the helicopter more dynamically to maintain a clear view

that the crew will undertake, but also the most enjoyable and rewarding.

The camera work on a pursuit is important, and the front seat observer needs to ensure not only that all sensors are in focus and set at a usable level of zoom. They also need on-screen imagery that captures the evidence required while giving ground resources commanding the pursuit as clear an overview of the activities and risks occurring on the ground. In daylight the majority of the pursuit will be filmed using the daylight (EOW) sensor with spotter scope (EON), which is used to gain detail such as driver and passenger descriptions as and when appropriate. The IR sensor may only be used should the driver decamp and a search for suspects become the priority. At night the situation is different: in well-lit areas the EOW can be used, with manual control opening up the iris to let in more light and generate a workable picture on screen, while the LL spotter scope (LL EON) can be used in almost complete darkness to read number plates and gain evidential details. The majority of activity, however, is on the IR sensor, which gives a clear black and white image on screen showing exactly what is happening in complete darkness. The TFO ensures that the blend of imagery is such that they are not too tight on the suspect vehicle so as to lose situational awareness but not too zoomed out so as to lose evidential detail. This balance is fine, and a competent camera operator and tactical commander will work between them to ensure that the camera work is first class.

Closing the pursuit

The closure of the pursuit is very much in the hands of ground resources and the driver of the suspect vehicle; it is highly unusual for a pursuit to last longer than the endurance of the police helicopter, although this does occasionally happen. The vast majority of pursuits result in either some form of crash and/or decamp, when the driver decides to abandon the vehicle and make a run for it. The presence of the helicopter increases the chances of this part of the pursuit being successful, and the crew can seamlessly switch from following a vehicle to tracking the driver and sometimes multiple passengers as they make a run for it. The role of the helicopter crew is to maintain visual contact with as many of the suspects as possible, focusing on the driver who has most likely committed the major offences, and then guiding ground resources including police dogs to search for, locate and arrest them. The pursuit becomes a suspect search, and the crew adapt roles accordingly to ensure that the chances of success are maximised.

Public order

The origins of air support lie within the public order arena: police helicopters were introduced to provide the aerial overview that was required to effectively police large-scale public order events. Other air support tactics grew out of the desire by various pilots and TFOs to maximise the operational benefits that were available through access to air support. In most other tactical roles, the helicopter crew members are effectively working to deliver a specialist capability that supports ground resources. They are the experts, and they determine what to do and how to do it when engaged on task, using the role equipment to its maximum benefit. In public order tasking the situation is slightly different, as the helicopter effectively becomes a tactical option that is available on behalf of the command team, deployed when the command team choose to deliver the imagery they need to police the event.

Command team

Police command at any public order or other major incident is based around a gold, silver and bronze team structure:

Gold sets the strategy for policing the event and is in overall command of the event.

Silver determines the tactics to be used to achieve the strategy set by Gold. Silver has command of all the resources and deploys them as required via various Bronze commanders.

Bronze retains command over a specific tactical option, area of the event or group of

tactics. Bronze commanders are in charge of resources within a specific sector, such as reserves, traffic, criminal investigation, mounted, marine or air support.

The command team meets many times in advance of any public order event as plans are drafted, tactical options are discussed and resource levels are agreed. The planning process varies according to the scale, complexity and risk of the event. The role of air support can be determined in advance, in terms of the number of helicopters, times of deployments, capabilities to be made available and the amount of flying required. Air support staff provide tactical advice during the planning phase of any public order operation, and ensure that the correct resources are in place and ready at the times and locations needed.

In London major public order events are commanded from a special operations room (SOR), a specialist dedicated control room; this uses the call sign 'GT'. In other parts of the country forces have their own command centres either adjacent to or contained within their force command and control structure. These event control rooms can contain a multitude of different command pods, each resourced by specialist control room staff and managing the deployment of resources within that area. Events are typically managed via a range of radio talk-groups: the command channel is used for all command team members to communicate effectively and working

ABOVE The SOR with the call sign 'GT' is used to coordinate resources and police any major event in London, from public order demonstrations to state visits. The air support unit embeds a liaison officer as Bronze air support, tactical adviser to Gold and Silver. This photo shows one of the many video walls that routinely display the helicopter's downlinked pictures.

channels are used for deployment of resources and management of the event within a specific footprint. The size and scale of the event will determine the number of working channels needed and the complexity of the control room set-up. Gold and Silver are typically based within the control room, with various Bronze commanders out in the event commanding their specific resources as directed by Silver.

In London the ASU is routinely employed as a key tactical option for a range of major public order events, such as protest marches, static demonstrations, state events, and ceremonial and royal occasions. One of the most important elements of a public order tasking is the downlinking of live video pictures from the helicopter into the control room and local receivers. The crew takes on a number of different roles depending upon the type of event, what is going on and how it is progressing. During the pre-event build-up these can begin with security searches and checks, during which the crew works to methodically cover the routes, locations and search areas defined by the police security coordinator and the PolSA. The timings of these activities will be precise, and the helicopter crew will be given times to be on task (referred to as ground assigned) and a window of time within which to complete their searches.

As the public start to arrive, the role may switch to providing overview imagery and situational updates for the command team who are monitoring policing plans and the deployment of their resources. During the event itself the ASU may be required to provide periodic situational updates, meaning they come and go as required, or they may be requested to provide a continuous presence, which requires two helicopters to piggyback each other. The first helicopter is on scene with a specific flight endurance of, say, 120 minutes, while the crew of the second helicopter will agree a time to be on scene for a handover, allowing the first sufficient time to return to base for a refuel and land with at least MLA. This all takes careful coordination, as things can change: the fuel burn rate may be greater than planned, bringing forwards the handover, or indeed may be better than anticipated, therefore pushing it back. The coordination of air support resources is managed by the ASU liaison officer, an ASU manager within SOR.

ASU liaison role

As mentioned earlier, an event is policed by a command team that comprises a number of Bronze commanders. One of these will effectively be 'Bronze Air', a role undertaken by an ASU manager, typically a sergeant or inspector. The ASU liaison officer will attend the briefings at SOR and be located within the command centre for the duration of the event, often for in excess of 12 hours. During this time they manage ASU deployment, provide tactical advice to Silver and Gold, and ensure that the ASU is able to react quickly to emerging incidents by pre-empting deployments – based upon radio traffic and updates from various sectors. The ASU liaison has a control pod within SOR and is able to monitor the electronic message switching system, control the downlinked imagery, and communicate on a multitude of different radio channels. This is all achieved through a workstation with multiple monitors and an integrated command and control system.

The video imagery from the helicopter is displayed on large video walls throughout the main SOR control room and also in special Silver and Gold suites and meeting rooms. At any given time, dozens of senior officers from police, fire, ambulance and other partners can be viewing the imagery and using it for their own purposes. When downlinking live imagery that is

BELOW The ASU liaison pod in SOR from where the air support liaison officer can manage the helicopter, providing downlinked imagery and radios as well as linking into the events command and control computer system.

being viewed on the video walls it is important that the crew understand the impact of sudden camera movements upon their viewing audience. In normal air support work the main purpose of the camera sensor is to be used as a tool for the crew, so the camera moves around rapidly changing sensor, field of view and focus as the front seat TFO exploits the scene, either looking for a suspect or ensuring that a fast-moving vehicle is kept on screen during a pursuit. Public order is inherently different, in that the movement of the camera needs to be slow and considered: those viewing the imagery are on the ground rather than in the air. The crew needs to consider what the command team will need to see and give them quality imagery for long enough that they are able to process it, make informed command decisions and ensure tactical deployments are working as planned. The screens in SOR can be several metres high and wide, and sudden camera movements can cause those on the ground to feel ill and even fall off their chairs.

The use of technology is increasing rapidly, and live downlinked imagery can be viewed by Bronze commanders on the streets via hand-held downlink receivers, public order carriers equipped for downlink and smart devices with internet access that can view a secure downlink video stream. The helicopter crew at a public order event is acutely aware of how important imagery is and how it will be used, so the use of the camera is more like that of a TV broadcast than it is in normal helicopter video operations.

All helicopter deployments are managed through the ASU liaison officer rather than the usual command and control set-up. The liaison officer attends the regular Silver meetings and receives situational updates throughout the event from various members of the command team. Event policing plans can change many times as the event unfolds, and it is important that the crew remains vigilant and responsive: their performance and work is visible to everyone, and the command team of senior officers will judge the ASU based upon how it performs in these one-off events rather than in day-to-day policing. It has often been said that when the ASU does not turn up at a garden search nothing happens, but when it fails to turn up and deliver at a public order event the

phone never stops ringing, and the wrath of frustrated senior officers is felt. For this reason alone, the command of the ASU is handed to SOR, coordinated by an ASU manager as liaison officer, and business as usual policing becomes a secondary consideration. It is not that the crew will not undertake routine tasks such as searches or pursuits; it is just that when engaged on public order events they will only do so with the consent of Silver and as directed by the ASU liaison officer.

Eagle One

When the EC145 was being role fitted, the design team wanted to provide the ability to carry a commanding officer at a major, critical or public order incident and therefore to provide them with their own dedicated imagery and communications. The ASU had always had the ability to land and pick up a senior officer should this be required. An example of this occurred at the large fire that took place following the Buncefield oil storage facility explosion in December 2005. When a senior fire officer was flown by the ASU over the fire, the crew reported that he turned to them and said 'We're going to need more fire engines' as he surveyed the scale of the incident. The benefits to any commanding officer of an aerial overview are clear, and although live camera imagery can be downlinked, the wide angle perspective gained by the view from a helicopter window

BELOW A QUAD screen image showing all sensors during a public order event in north London. The tactical commander is able to review all available imagery as well as the mapping and use this to maintain situational awareness.

ABOVE Public order events can take place day or night, and the IR imagery is sometimes the only imagery the ground commander has to work from. In this picture public order resources are advancing to clear a residential street.

cannot be underestimated. With this in mind, the Silver command position was developed on the EC145, which came into service in July 2007. In reality Silver was never going to fly, as they were needed on the ground and could not command an event from the relative isolation of a helicopter cockpit some 1,500ft above the event; but the name seemed to work and it stuck. It took several years of EC145 operational use before the vision of this Silver command position was finally realised.

Public order command teams are drawn from a group of highly trained and experienced public order commanders known as the public order cadre. These officers work in teams; they grow very familiar with each other and the various ways of working that each Gold and Silver have. The idea was to create a new role that sat outside the ground-based command team, to be filled by a member of the cadre at chief inspector or superintendent level. This role would have the call sign 'Eagle One' (or 'Eagle Two' if there were two helicopters) and would fly with the crew of the helicopter throughout the event. Selected for the role by Silver, they would not have specific policing responsibilities on the ground but would be fully aware of the tactical plan and provide detailed advice and support to Silver on the progress or otherwise of the plan. The role was popular with many senior officers but not all; the early days saw a number of cadre officers become extremely air sick, which somewhat detracted from their role. Others loved it, however, and became very effective and efficient additions to the crew – having learned how to operate the technology and understanding the finer details of safe flight operations. The Eagle One role continues to this day, and is used at major policing events where the presence in the air of a command team member who can communicate peer to peer with Silver is deemed useful.

A phrase frequently used at Lippitts Hill was 'public order pays the bills'; crews often preferred the more dynamic and exciting police

RIGHT Not all public order events involve the potential for disorder. In this picture, taken during a UK visit by the Pope in 2010, the air support imagery was used to ensure public safety during the build-up to and after the event.

work provided by searching and pursuits and found public order to be slow and monotonous. However, when things went wrong at a public order event and disorder broke out the crew earned their money, providing a valuable and dynamic response to the unfolding events. They were able to switch effortlessly from the slow and considered downlinked imagery that was needed when things are going well to tracking suspects, capturing evidence and supporting resources on the ground that were deployed to regain control. Public order and event policing may not be the most exciting use

ABOVE LEFT Notting Hill Carnival, which takes place in London every August, attracts around one million visitors and presents a range of public order challenges for the police. This image shows the density of crowds in what are normally residential streets.

ABOVE This image shows the coordination of an advance by mounted police officers during a public order clearance.

of air support, but it carries the greatest risk to the public, police and reputations, and the presence of air support is considered essential in order to ensure the event passes off safely for all involved.

LEFT Public safety is a policing priority, and this image shows the view of the tactical commander's workstation during the Oxford v. Cambridge annual boat race in London. Crowd density on the river banks can be an issue, and here the crew is looking for signs of crushing and providing a situational overview for the command team.

ABOVE The term 'public order' is used to refer to a wide range of policing and public safety events and not just demonstrations. In addition to state and ceremonial occasions, London, as the capital city, does see its fair share of demonstrations and protests, the vast majority of which are peaceful. Here, a protest group is seen on the tactical commander's display marching near Hyde Park.

RIGHT Public order policing of a sporting event, in this case the Tottenham v. Arsenal local football derby. Public order officers are escorting the Arsenal (away) fans to the Tottenham stadium.

ABOVE The London Marathon is another annual public order event where the presence of the helicopter is for security and public safety. It provides live imagery for police command teams. The runners are crossing Tower Bridge.

LEFT A public order event in Westminster is filmed from the police helicopter and downlinked live into the control room.

Using infrared

The development of thermal imagery (or IR) over the last few decades has revolutionised police air support capability. Being able to search in complete darkness using a thermal imager is exceptionally valuable, given that over 60% of air support activity is search related. The increasing use of drones by police has driven the deployment of small IR sensors that can now be fitted to mid-range drones. These sensors are nowhere near as high resolution and capable as the cooled large format IR sensors fitted to police aircraft turrets, but they deliver an excellent capability for night-time search activity. The use and interpretation of IR is a complicated subject and we will not seek to delve too deeply into the science behind it here, instead focusing more upon its practical application by the TFO operating the sensor.

Thermal imagery in basic terms is the interpretation of a scene based upon the IR return of the objects within it. The IR sensor detects IR energy, or heat in simple terms, and displays that heat picture in a format that the human brain can interpret. The result is a high-resolution image displayed on screen that appears to show the exact scene in greyscale. The detail provided by modern IR sensors is exceptional, and it is literally possible to create an accurate picture of what is happening within any given scene in complete darkness. The problem is that what you are looking at is heat and not light, and as such things are visible in IR that are not visible to the naked eye in daylight or under a searchlight and things that are visible in light may not appear the same in IR. It is important, therefore, that TFOs understand the basic scientific principles behind the IR imagery they are using and learn to interpret and exploit the imagery on screen to maximum effect.

The electromagnetic spectrum

The electromagnetic spectrum ranges from gamma rays at one end of the spectrum through to radio waves, such as UHF, FM and AM. The tiny section in the middle of the spectrum known as visible light is what we are able to see with the naked eye. Once this visible light disappears, we are plunged into darkness, and we are no longer able to see with the same degree of efficiency. IR energy is generated by hot objects such as the sun and is harmless to humans. By using technology it is possible to convert the invisible IR energy into a picture that can be understood and interpreted by the naked eye.

The sun is the prime source of IR energy, but as IR is effectively heat, other hot objects can impact upon an IR picture. All matter responds individually to IR radiation, and this assists us when looking at interpreting IR imagery. All objects either reflect or absorb and then emit IR radiation; the extent to which they do so is determined by their emissivity properties. Radiation comes from two sources. The first is active sources, such as human beings who emit radiation as they get hotter. This is important because suspects who have run from police will be hot. The second is passive sources, such as the sun: radiation from the sun hits objects on earth and is then either reflected or absorbed and then transmitted by the object. It is by understanding these basic principles that it is possible to see how the IR sensor works. In very basic terms, it features a supercooled forward-looking array that is made up of millions of tiny pixels. Just like the pixels that make up any digital image, these are each capable of detecting IR radiation (heat). The smaller and higher the number of these pixels the greater the resolution present within the image, and the more detailed the IR picture displayed on screen.

BELOW A diagram that explains the electromagnetic spectrum.

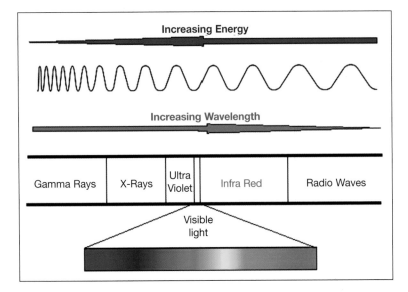

Increasing Energy

Increasing Wavelength

| Gamma Rays | X-Rays | Ultra Violet | Infra Red | Radio Waves |

Visible light

RIGHT The clarity of the IR imagery turns night into day for the helicopter crew, and allows them and ground commanders to see clearly what is happening any time of the day or night. The interpretation of the IR image can be challenging, though, as things do not always appear in IR as you might expect.

In very simple terms, the IR sensor looks at the contrasting levels of heat emitted by objects present within a scene. The greater the contrast between two objects the more clearly they are defined on screen. This means, for example, that when looking at a suspect on screen in IR the sensor will be able to detect the differing thermal characteristics of the components of their clothing. A zip or an embroidered logo will have a different thermal characteristic to the remainder of the cloth. However, different fabric colours or patterns will not necessarily show up in an IR image, and therefore a striped or patterned top may appear to be plain. It is important that the TFOs learn to carefully interpret imagery on screen, which may look very different in real life.

Using IR imagery operationally

In order to get the best performance from the IR sensor the TFO needs to calibrate the IR while looking at the scene that is to be searched. Calibration is achieved by a simple press of the calibration (IR CAL) button on the hand controller. The IR sensor is instructed to look at the hottest and coldest elements present within the scene and apply black to one and white to the other. The thermal returns of objects in between the hottest and coldest items present are then represented as shades of grey. The result is a greyscale picture that is exceptionally realistic. It is important to select a field of view that will be helpful for any given search or activity, and when the scene changes dramatically it is necessary to re-calibrate the IR sensor. This occurs in reality only when moving from an urban setting to a rural setting or from land to a search of open water. The TFO can select to work with black as hot and white as cold, or vice versa. This preference is changed by simply pressing the polarity (POL) button. With early generation IR sensors, the TFOs would usually have a preference, and

ABOVE In this image a man is threatening police officers with a scaffolding pole in the rear gardens of a residential street. The male is in the dark and standing on top of a garden shed. The crew illuminated the rear garden using the searchlight for officer safety and then used the IR imagery to obtain a detailed view of the man's actions. In this case the man, who was believed to be suffering from mental health issues, was safely removed from the roof and taken to a place of safety.

would choose to work 'black hot' or 'white hot' almost exclusively. With the latest generation of IR sensors, however, TFOs tend to flip between the two formats throughout the search. The contrast between items present may show up better in one format than the other.

IR imagery is a huge operational benefit when searching for suspects, as it literally allows the

RIGHT This image shows the use of black hot IR imagery.

FAR RIGHT This image shows the view of the same scene in white hot IR imagery. The TFOs routinely look at a given scene in both formats, as sometimes details show up in one but aren't visible in the other.

helicopter crew to see in the dark – and the resolution presented on screen is exceptional. Some suspects show up clearly in IR even though they may think they are in complete darkness and out of sight. It is very common to find suspects lying on rooftops in the dark, thinking that officers walking past will never find them. The TFOs look for anything that does not seem as if it should be there. If working 'white hot', and a fence panel is showing signs of being hotter at one point than another, indicated by a dark patch, the TFO will ask why this is. It is possible that one part could be rotten and damp, meaning that it has a different thermal characteristic. Alternatively, there could be a suspect hiding behind it, and transferring body heat into the panel. From this

example alone it is possible to see how the IR sensor gives TFOs a significant operational advantage. The naked eye could not see through a wooden fence yet the IR sensor can.

It is clear that IR appears differently on screen to conventional video pictures, and this can cause problems if a TFO is not careful. An example of this is so-called thermal scarring, which is where an object that was previously present has moved, leaving behind a scar that is still visible on screen. An example would be a car that has shielded the ground beneath it from the sun during the day, causing it to absorb less IR from the sun; when the car leaves, the ground beneath it will appear cooler than its surroundings. The result will be a shape on the IR image that appears to be the vehicle that was previously parked there. Another example of this is the thermal residual effect, when an object transfers heat to its surroundings, and this residual heat is then visible on IR. A frequent example would be when a suspect has been hiding in a specific location, perhaps lying on a flat rooftop, and their shape remains on IR imagery after they have gone. However, IR can be very useful if suspects remove clothing or throw away items that may be potential evidence. The residual heat in these items means they show up clearly in IR even if in the darkness they might blend into their surroundings; this makes evidential recovery easier.

The science behind IR is complex, but a basic understanding of how it works and what it represents on screen enables the police helicopter crew to use it to maximum effect operationally. The ability to see in the dark presents the crew with a huge advantage when policing at night. Whatever role they are undertaking, IR imagery can be used alongside conventional video imagery and LL enhanced imagery to achieve operational success.

RIGHT This suspect clearly shows up in IR even though he thought the complete darkness meant he was hidden from view.

01:24:14

RIGHT The small black circle at the bottom left of this shed door was sufficient to give away the presence inside of an aggravated burglary suspect.

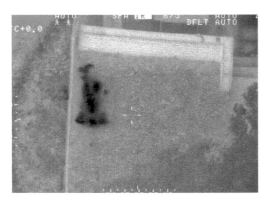

ABOVE This image shows the residual heat from a suspect who was lying in a rear garden to hide from police and then moved. TFOs need to be careful to understand what they are looking at, as this person-shaped heat source could have been confused for the actual suspect.

RIGHT This image of an airport shows a thermal scar where an aircraft was once parked. Sometimes this can be caused by the regular positioning of a vehicle or object, which changes the weathering of the tarmac or concrete at that point and gives it a different thermal characteristic. In this case the white dot (white is hot) shows where the aircraft's external power unit was placed; the aircraft has just moved.

BELOW The thermal residue of this suspect can be clearly seen: this is where he was lying down.

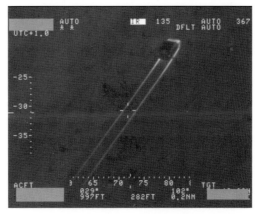

LEFT This IR image shows the space where a vehicle was once parked. In this picture black is hot and the ground beneath the car is warmer than its surroundings. This may be because it was shielded from the cold or because a running engine has heated up the tarmac. TFOs learn to question and interpret imagery.

LEFT In this image the skid marks from a vehicle show up in IR, indicating the point at which the brakes were applied and generated additional heat. This can appear different to physical skid marks on the tarmac that are caused when the wheels lock and tyre rubber is deposited.

LEFT In this image the suspect has thrown an item of clothing into a rear garden. It shows up clearly in IR and remains hotter than its surroundings for some time. This allows the recovery of evidence even in complete darkness.

Chapter Six

Police helicopter in action

Two burglary suspects chased from a suburban London home ran into gardens to make their escape. Four suspects in a vehicle pursued by police after failing to stop drove the wrong way around the M25. Air support was there to manage risk and bring the suspects to justice. These case studies take you along with the crew as they use air support to fight crime.

OPPOSITE Local police officers supported by a dog and handler arrest a pair of suspects. They were spotted by the helicopter crew hiding in a suburban garden shed. Helicopter crews and dog handlers have a close working relationship and routinely train together to enhance their understanding of one another's roles.

Case study 1 – Criminal suspect search

Date: 23 March 2010
Time: 2330hrs
Location: London Geographia page 120
Comms: SX Despatch 1

The crew is scrambled to search for burglary suspects, who were chased by police and lost as they turned into an alleyway leading to rear gardens in Hendon, north London. Local officers have a containment in place. Air support is requested to attend and search rear gardens between three Hendon roads. The location is south-west and around five minutes' flying time from the base.

Last sighting

2330hrs Scramble

The pilot and front seat TFO scramble from the crew ready room and strap themselves into the helicopter, which is on the helipad and ready to fly. The tactical commander meanwhile obtains as much detail as possible from the operations room and monitors the local radio talk-group on a personal radio for updates that might impact the tasking. After leaving the ready room, air-crew helmet on, the tactical commander takes up position outside the rotor disc at the front of the helicopter. The pilot indicates engine 1 to start and the tactical commander acknowledges this, moving towards the port side of the helicopter as the engine bursts into life and the main and tail rotors begin to rotate. Soon the engine is safely started, and the process is then repeated for engine 2. A thumbs up from the pilot indicates that the tactical commander should walk under the spinning rotor disc and approach the helicopter. After a short safety walk round they enter the cabin, plug their helmet into the intercom and strap in, announcing that they are secure and ready. Meanwhile, the front seat TFO is powering on and configuring the police role equipment, starting at the back of the centre console and working forwards, making sure everything is working as it should. With the crew strapped in and ready, the pre-take-off checks are completed and the helicopter is ready to lift; this process takes around four minutes.

2334hrs Take-off

The helicopter is airborne and heading west towards the target location. The pilot speaks to ATC and gains clearance to fly directly to square 120 while the tactical commander is selecting the working talk-group 'SXDesp1' and entering the map reference into the mapping system. The front seat TFO is preparing the camera, ensuring the IR has cooled and is calibrated and that the IMU has aligned and the system is operating in auto mode. When everything is ready the tactical commander contacts the local control room by radio and announces the presence and estimated time of arrival of the helicopter. It is clear to the crew that they are to search enclosed rear gardens after suspects were seen by police to run along an alleyway towards the gardens and disappear from sight. Priorities at this time are to get the recorders started, camera ready and pointed at the target as they transit, and when on scene to quickly identify the search area. The tactical commander selects the target area on the map and presses the SEND POS button on screen to send the target coordinates to the camera system. A quick press of the slew button on the hand controller, and the camera is pointing at the target from many miles away. **PIC** **1**

2338hrs Arrival on scene

The arrival of the helicopter on scene is fairly obvious to the officers on the ground; there is no mistaking the sound of an orbiting helicopter. As the crew arrive, communication with the contact officer on scene is established by the tactical commander, video recorders are recording, the digital downlink is transmitting and the front seat TFO has the camera calibrated and ready to work. The search area is confirmed by radio as being the rear gardens enclosed between three streets in an elongated triangle shape. This is clearly visible out of the window as a dark shape outlined by the yellow and white street lights. The tactical commander, front seat TFO and pilot all agree that they are looking at the same visual location and that the camera is on the target area. **PICS** **2** & **3**

2340hrs Starting to search

The tactical commander talks by radio to the officer coordinating the ground activities and confirms the description of the suspects, direction of travel, last known location and any

Every alleyway and canyon between houses must be searched methodically

Security light just come on

Shed door ajar?

other information. The search of enclosed rear gardens involves the pilot initially putting the helicopter in a right-hand orbit around the target area, allowing the front seat TFO to look at the search area with the IR sensor looking for obvious signs of movement and heat sources. The police vehicles are identified, as are the officers providing the containment. The crew then start a systematic search, working from the area the suspects were last sighted in and following guidance from officers, who believe the suspects are in a specific set of gardens. **PIC 4**

2340hrs   Heat sources investigated

The crew identifies a warm patch on a fence panel in one of the gardens, and the garden lights clearly show up as hot spots. Experience tells them that a suspect leaning on a fence panel can cause it to warm up, but so can dampness or even heat reflected from a garden light. They continue their orbit to look at the back of the fence panel and nothing obvious shows up. The crew feel it is unlikely to be a suspect. **PIC 5**

2341hrs   Search continues

The search progresses as the helicopter orbits, and each garden is searched from all angles using the IR sensor in wide angle and narrow angle fields of view. The alleyways between houses and gaps between buildings are ideal hiding places, and the pilot positions the helicopter to allow the front seat TFO to look directly into each in turn. All of them appear clear of suspects. The containment is good and was in position quickly, with officers located in all surrounding streets, so the chances of the suspects escaping are low. It's almost midnight and there are few other people around; most residents are in bed. **PIC 6**

2342hrs   Search continues

Suddenly an officer on the ground says that a security light has come on in one of the gardens; this could be the suspects breaking cover. The house number is given from the ground and quickly identified by the tactical commander using the mapping system building outline layer. Quickly the target is selected, SEND POS button pressed and the front seat TFO advised that they can again slew the camera to the target. The house in question appears on screen but the garden is clear: it's a false alarm, possibly an animal or the occupier has activated the light. It is very common for the noise from the helicopter to draw residents out of their houses and into their gardens. Although security light activations can be red herrings, they must all be investigated. **PIC 7**

2349hrs Keep searching

The search continues relentlessly, and every potential heat source is methodically checked and eliminated. The gardens are full of sheds, trampolines and shrubs, which all provide potential hiding places especially in the dark. The IR camera doesn't care about the darkness, though, and the imagery presented to the crew is as clear as daylight, just black and white. Water features can appear hotter than their surroundings at night, as water absorbs IR radiation all day and emits it at night. Compost heaps and bins also appear warm, as the decomposition of vegetation generates heat. All these things can look hot compared with their surroundings, and the crew must investigate them all to ensure they are not a suspect. Suddenly the front seat TFO sees a shed in one of the gardens where the door appears to be open. **PICS 8 & 9**

2349hrs Found them

A closer look at the shed reveals movement inside, and the door is pulled shut. It is almost certain that the suspects have been found, and the tactical commander immediately communicates this to ground resources while simultaneously identifying the exact location of the shed. The risks associated with searching for suspects can be significantly managed and reduced by observing their actions from the air and using a police dog to complete the initial approach and confirm the suspects' presence. There is no intelligence that the suspects are armed in this case, but they could be, and this unknown factor adds to the risk to officers as the suspects are clearly trying to evade capture. **PIC 10**

2350hrs Maintain observation and coordinate response

The priorities for the crew are now to maintain observation on the shed in case the suspects make a move, and to coordinate the ground units so they can move in, search the shed and make arrests. The tactical commander works out exactly where the shed is and identifies the best access route for the dog handler. The officers are directed to an access road between two houses and advised to await the arrival of the dog handler and police dog. **PICS 11 & 12**

2352hrs Dog handler arrives

The tactical commander takes control of the situation, as although the local officers want to move in and arrest the suspects, the dog handler is very close and within a couple of minutes can deploy the dog to search the shed. The officers and dog are guided down the access road, then advised to turn left past a skip: they will find the shed on the left-hand side. **PIC 13**

2353hrs The dog is deployed

The dog is deployed ahead of the officers to search around the shed, then the shed door is opened and the dog goes in. It is not possible to hear the barking or indeed what is said on the ground, so the crew can only imagine how the officers and suspects react when they come face to face with each other, especially with the assistance of the dog section.
The local radio bursts into life with the short and simple message 'Got them, two detained.'
PICS 14 & 15

2353hrs Two arrested

A short while later two suspects are removed from the shed and secured in handcuffs: two arrests and an excellent job all round. Just 23 minutes has elapsed from the call for air support to the arrest of the suspects; the value and efficiency of air support is clear for all to see. One wonders if the helicopter had not been available how long the search would have taken and what the result would have been. **PIC 16**

Case study 2 – pursuit and search

Date: 4 December 2009
Time: 0216hrs
Location: London Geographia page 147
Comms: PMPSINTOP

The helicopter is just finishing a missing person search when the radio bursts into life on the Interoperability channel. A vehicle is failing to stop for police on the A13 eastbound towards the M25. The crew accept the tasking, and advise the radio operator that the helicopter is on its way and will be on task in about four minutes.

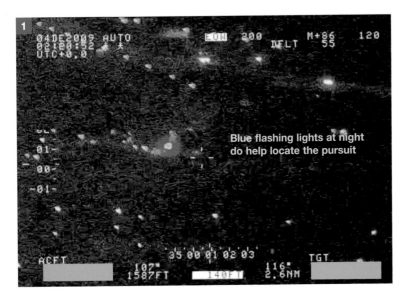

Blue flashing lights at night do help locate the pursuit

0216hrs **Work in the cockpit starts to ramp up**

The pilot obtains clearance to square 147 and advises ATC that the task is a pursuit with the location moving east. The front seat TFO looks at the map book on their knee, and using local knowledge immediately guides the pilot towards page 147; the helicopter is already flying east as the pilot knows the A13 well. The tactical commander selects the target on the mapping using the cursor and then sends the position to the camera; a quick slew later and the camera is pointing towards the target area. Video recorders are stopped at the end of the last task and immediately restarted to ensure a new recording chapter is created, the radios are switched to use TAC2 and the front seat TFO confirms that TAC2 is being recorded. The crew are working hard, listening to the radio for updates and hoping that the pursuit continues long enough for the helicopter to arrive on scene. This way the risks are lowered and the chances of catching any suspects are high.

0220hrs **Pursuit is spotted**

The helicopter is flying at 1,500ft above ground and is still 2.3 miles away from the blue lights that appear to be on the A13 heading east. The pursuit is authorised on the radio and the driver of the primary vehicle, an advanced pursuit trained driver, is assigned as ground commander. Until the helicopter is overhead and established the crew do not declare that they have the pursuit on screen. It is very easy to chase shadows, and there are blue lights on emergency vehicles all over London at night; they could be following a response car or ambulance going to a completely different call. The pursuit is approaching Junction 30/31 of the M25, the roundabout where vehicles can leave the A13 and join the M25, when the commentary from the ground commander confirms that the vehicle is off at the junction and heading towards the M25. PIC **1**

0222hrs On-scene

The direction of travel is now critical, and just as the suspect vehicle enters the roundabout the helicopter arrives overhead. The camera, now in IR mode, displays on screen a clear view of the suspect vehicle. The front seat TFO uses various fields of view and daylight camera to confirm that the correct vehicle is being observed, the pilot looks at the screen and out of the window, ensuring that they too are following the correct vehicle, and all this is confirmed by the tactical commander; the communication via the cockpit intercom is busy but precise and measured. The tactical commander switches the mapping screen to camera position mode and a line is drawn on screen with a cross hair on the map, pointing exactly to where the camera cross hair is located, at the position of the vehicle. The mapping displays the GPS speed of the vehicle and the name of the road, there is a quick check with the crew, 'everybody happy?', and then the tactical commander announces on the radio that they will take over the commentary. PIC **2**

0223hrs Helicopter crew take over

It is 0223hrs and the helicopter is now in control; there is often a tangible sigh of relief as the ground commander, who retains overall control, can drop back slightly and focus on the driving and the conduct of the suspect vehicle rather than where they are and what to say over the radio. Despite the adrenaline pumping

the on task activity in the cockpit is very calm: every crew member is doing their job and communication between them is continuous, to ensure everyone is aware of what is happening. The tactical commander gives updates to the radio controller call sign 'MP' of the vehicle's exact location, estimated speed, direction of travel and any pertinent information about other traffic, offences committed and road conditions. Once the vehicle is on screen it is very uncommon for the crew to lose contact with it, and the chances of bringing the pursuit to a successful ending increases as the risks associated with doing so reduce. The vehicle is on the M25 heading clockwise towards the Queen Elizabeth II bridge. PIC **3**

0223hrs Madness on the M25
Suddenly the vehicle slows and stops. It's the M25 – there's nowhere to go, surely; they're not going to decamp from the vehicle here? The vehicle comes to a complete stop, as do the police vehicles and other M25 traffic. Thankfully its 2.23am, so there's not much traffic. The vehicle then does a complete 180° turn and starts to reverse away from the pursuing police vehicles. It is now slowly driving backwards along the M25. Then it comes to a stop again. The police vehicles stay put: they don't want to move too quickly; the risks are high as the traffic behind them is still potentially travelling at speed. The vehicle then accelerates towards them and drives between the two police cars: it is travelling the wrong way along the M25 back towards Junction 30/31. PICS **4**, **5** & **6**

0223hrs M25 the wrong way!
The tactical commander immediately terminates the pursuit: the risks are too high, and there is no way that the police vehicles can continue their pursuit the wrong way along the M25, potentially facing at speed oncoming drivers who are not expecting them. There is nothing they can do about the suspect vehicle, so the helicopter is now the only police asset managing the pursuit. The tactical commander tries to coordinate other police officers to close

off the M25 at the junction and prevent any further traffic from travelling towards them. Thankfully the traffic is light and mainly heavy goods vehicles, the drivers of which immediately slow down. PIC **7**

0224hrs Decamp

Thankfully, the vehicle leaves the M25 via the on slip road and approaches the roundabout again, travelling the wrong way round the roundabout and taking the exit back towards London. The helicopter crew, who have maintained their view of the vehicle throughout this exhibition of extraordinarily dangerous driving, have continued with their commentary and are trying to talk police vehicles back in behind the suspect vehicle. Some 30 seconds later, however, the vehicle suddenly veers to the right and crashes into the grass verge before stopping. All four doors open and the four occupants all climb over a fence into a housing estate. 'Decamp', explains the tactical commander, and the crew activity now changes from following a vehicle to trying to follow four suspects – and at the same time coordinate the ground resources to contain them within the housing estate. Suddenly the relative calm and controlled environment in the cockpit is replaced by feverish communications and activities. The tactical commander identifies the local radio channel and starts to manage two radios in order to ensure that the original pursuit vehicles and local resources are all aware of the location of the decamp and the actions of the suspects. PICS **8** & **9**

0240hrs Containment and suspect 1 arrested

The suspects split up, with two heading further into back gardens, one going to ground quickly and a fourth slipping from sight. The tactical commander decides they will stick with the two who are moving through the gardens. The location of the third is noted and they will come back for the fourth. The suspects move quickly through the gardens, hopping over fences and slipping behind sheds and garages, then suddenly they vanish from view. PICS **10** & **11**

What's this? One of the original suspects has hidden in a garden outside of the containment. Crew note the position.

Dog locates suspect and arrest is made safely and effectively.

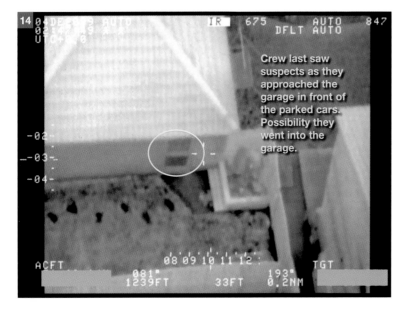

Crew last saw suspects as they approached the garage in front of the parked cars. Possibility they went into the garage.

0242hrs **Search for suspects 2 and 3**

The priority shifts to coordinating the ground response. The arrest of the first suspect who has not moved from their hiding place is number one priority, followed by using the dog to search for the further two suspects who the crew are certain can't have left the area in which they were last sighted. The dog handler and local officers are talked into the rear garden, where the suspect is hiding from view crouched in a corner behind garden furniture. The dog quickly finds the suspect and the first arrest is made: one in custody. **PICS 12 & 13**

2347hrs **Suspects 2 and 3 arrested**

The crew return to the rear gardens where the other two suspects were last seen. They continue to search and notice that there is a side door to a garage that was blind side to them as the suspects ran around the building's corner. The dog handler is talked round to the side garage door; the door is open and the dog enters the garage. A few moments later the call comes back over the radio that one suspect has been found. The tactical commander advises the officers that there may well be two suspects and that they should be cautious. Seconds later, the second suspect is arrested, hiding with their mate in the garage and assuming they were out of view and safe. **PICS 14 & 15**

2350hrs Out of fuel ... one final sweep before departing

The helicopter is almost out of fuel (a phrase used to advise ground officers that they need to return to base and land with MLA). The pilot states they have two minutes left on task, and the tactical commander orders one final sweep of the estate for the fourth suspect. Suddenly the crew spot a figure standing in a corner between a parked car and a garage. As they start to discuss what he is doing he walks forwards into the street. They are sure this is the final suspect, who continues to walk casually out into the street as though nothing has happened. A quick radio call later and one of the ground officers on the cordons moves into the street and approaches the suspect. The fourth arrest is made. Start to finish this incident lasts 35 minutes. With four suspects arrested, a stolen car recovered and nobody hurt, it clearly demonstrates the value of air support. The crew return to base knowing they have made a difference, but take no further part in the arrest of the suspect – not necessarily hearing whether there is a charge or an ultimate conviction. **PICS 16 & 17**

Officers have one suspect located but are warned by the crew that there might be two suspects in the garage.

Two suspects detained 0249hrs.

Helicopter almost out of fuel and crew complete one final search sweep as they advise ground units re 4th outstanding suspect.

Police officer approaches him and he is identified as 4th suspect and arrested.

Short case study 1

Suspect hiding under a van – This brazen burglary suspect thought they were safe hiding from view under a white vehicle parked on a drive in broad daylight. **1** The police van and police officers looking for him were parked next to the drive. The eagle-eyed crew spotted a sliver of heat under the vehicle **2**, and as they changed the viewing angle it became clear it was their suspect **3**. A short radio conversation later with the uniform officers standing nearby, and the suspect was promptly arrested. **4**

Short case study 2

Moped suspect arrested – This suspect riding a stolen moped was initially chased by police: he quickly left the highway and started driving through the walkways of a housing estate. The police vehicles terminated the pursuit owing to the unacceptable risks, but thankfully the helicopter was on scene and able to follow the suspect for several minutes – until he eventually stopped his moped, took off his helmet and strolled off into the estate. Sadly for him the helicopter crew had already arranged a welcoming party, and he was promptly arrested.

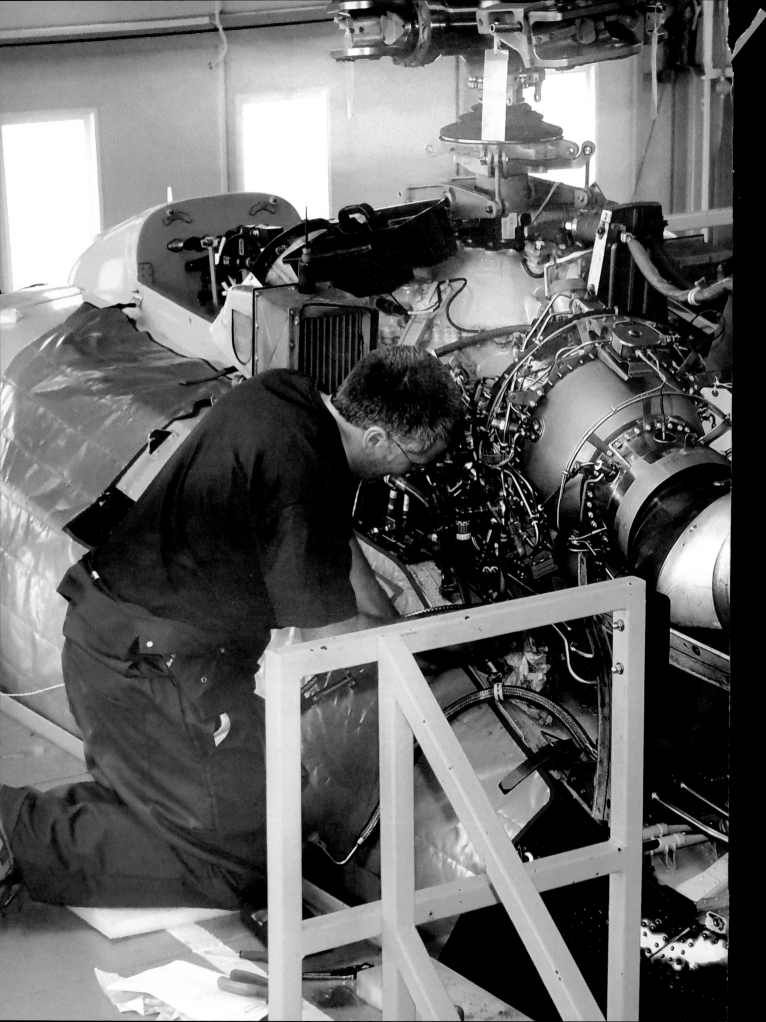

Chapter Seven

Behind the scenes

Effective air support requires a significant behind-the-scenes effort from the team of staff working hard to keep the unit flying. From the engineers who maintain the helicopters ready to fly, to the control room staff who filter hundreds of requests to ensure only those worthy of deployment reach the crew, everyone has an important role to play.

OPPOSITE Met engineers work on custom-built inspection platforms that give them safe and effective access to the engine and gearbox bays and the main rotor head.

Recruitment and selection

As air support evolved into a full-time operational command it became clear that the ASU needed to select officers with the right skills and attitude to be able to work in a very different policing environment. There are limited opportunities for TFO training as operational pressures mean that every flying hour used for training has to count. It is important to recruit officers who have the potential to become effective TFOs within the training time and resources available; therefore assessing aptitude for the role is crucial.

The process of applying for the ASU is known as selection, a description that was appropriated with pride from military and other existing recruitment processes. At the heart of the selection process are two different things: first, selecting an officer who will fit into a small team and work with minimal supervision; secondly, selecting an officer who demonstrates an aptitude for the role and can be trained within the time and resources available – in other words, identifying 'trainability'. TFO selection has evolved over the years, but the principles are the same and the key component parts are as follows:

Leadership and teamwork assessments – Candidates need to be able to work within a small team and must have the sort of personality that can take control when required without rubbing people up the wrong way. There is no space for egos, arrogance or people who simply can't get on with others.

Aptitude testing – This is completed through a number of video, map reading and listening exercises that are all related to the TFO role.

Flight-test – Candidates fly in the front left seat of a police helicopter, testing elements of navigation, situational awareness, listening and camera work. The purpose behind this element of the flight test is to assess if the candidate can take basic instruction and is able to put what they have been taught into practice. Experience indicates that the first thing that goes when a candidate is operating at capacity is hearing;

often candidates who are struggling to cope will miss key information and start to shut down their hearing and general situational awareness. During the flight test the candidate is asked more than once to bring the camera system out of its stow position and to steer it to look at a vehicle parked in a car park or to hold it steadily on top of a tall building. The aim is primarily to assess the candidate's ability to recall simple instruction regarding use of the camera but also to test their hand–eye coordination. The flight test can last about an hour, with a good candidate being able to navigate around the assessment easily within 50 to 60 minutes. Over the years it has been shown that candidates achieving above a specific mark are likely to become successful TFOs.

Helicopter maintenance

One of the main things that any police officer coming into air support needs to get to grips with is the difference between the maintenance requirements for your average car and those for a helicopter. All aircraft are maintained by approved maintenance and repair organisations (MROs), who employ engineering staff qualified and rated on the specific type of helicopters to be maintained. The maintenance schedule for any helicopter is defined by the manufacturer, with details of the intermediate and periodical maintenance inspections required along with part replacement to be completed at specific flying hours or calendar intervals. The MRO then produces a maintenance organisation exposition, which outlines in specific detail how they intend to achieve the maintenance requirements laid down by the manufacturer's schedule. Maintenance activity can be defined as daily actions, intermediate inspections or periodical inspections, these are typically required at intervals of 30, 50, 100, 400–500, 800–1,000 hours or annually, depending upon the rate of flying. Maintenance activity then escalates as the flying hours increase, with additional checks at intervals such as 1,600, 2,400, 5,000 hours and so on.

Police helicopters typically have high annual flying hours, and there are many occasions where NPAS-operated helicopters are world leaders in terms of the number of flying hours

LEFT One of the Met's EC145 helicopters in the south hangar at Lippitts Hill. The crane used to support the helicopter for the removal of its skids is visible in the background.

achieved on the specific helicopter or engine type. Average rates of annual flying on each NPAS airframe are around 900–1,000 hours, meaning that a ten-year-old helicopter can easily have completed 10,000 flying hours. There is no doubt that the maintenance provision within police air support is the very foundation of this operational activity. Without adequate maintenance provision the helicopters are simply not available to fly, and it doesn't matter how many pilots or TFOs there are: air support delivery is not possible without a helicopter to fly in. It is for this very reason that ASUs have always established a close relationship with their MRO, something NPAS continues today on a national basis across its fleet of 19 helicopters.

Helicopters are maintained under EASA Part 145 regulations and their continuous airworthiness under EASA Part M. Police helicopters are considered to be 'state aircraft' under EASA regulations, and as such in many ways EASA determines that it is up to the home regulator to determine their maintenance regime. In the UK it is the CAA that determines how state aircraft should be regulated and maintained. The CAA indicates that state aircraft should effectively be maintained exactly the same as any other aircraft, so that standards and safety are no lower. They issue British civil airworthiness regulations (BCAR), which govern the maintenance and continued airworthiness

of all state aircraft. They must be maintained in accordance with the manufacturers' requirements by personnel qualified and authorised on the type and within an MRO that holds approvals under BCAR A8-23 and A8-25 for maintenance and continued airworthiness. As these effectively mirror EASA regulations under Part 145 and Part M there is, on paper, little or no difference between the different maintenance regimes. One thing the police and emergency services must consider, however, is that if a helicopter is maintained only under BCAR then it may be difficult to sell the helicopter at end of service life back into the commercial aviation world.

Each helicopter is issued with a non-expiring certificate of airworthiness after a back to birth review of all maintenance carried out on it against the manufacturer's schedule. An engineer from an approved continuing airworthiness maintenance organisation (CAMO) undertakes the review and the Airworthiness Review Certificate (ARC) is issued. This ARC lasts for three years with a further interim review after the first and the second years, then another complete back-to-birth review every three years. This process ensures that the maintenance of the helicopter, right back to its factory build and delivery has been completed and documented correctly and that it is airworthy. In simple terms the maintenance of a helicopter can be outsourced to an approved

BELOW One of the Met's EC145 helicopters supported from the rotor head from a crane in the south hangar while the skids are removed. The maintenance schedule requires inspection of the cross beam assembly and mountings, for which removal is required. This picture was posted on social media and caused some complaints, as Health and Safety commentators suggested it was unsafe to prop up a 3.5-tonne helicopter on hydraulic stands. It was not clear in the photo that the aircraft was actually suspended from a crane.

MRO with approvals for that type of helicopter. It is the responsibility of the operator to ensure that the helicopter is airworthy, and to do this they must use an approved CAMO to direct, authorise and control all maintenance activity. This separates the actual maintenance activity from the continuing airworthiness review and the issue of an ARC.

The Met's ASU was one of only two ASUs in England and Wales to use its own in-house MRO; the other unit was the Essex Police ASU, which established Eastern Counties Police Maintenance and undertook maintenance activities for Essex and Suffolk ASUs. Both of these units were closed when NPAS was formed, and the maintenance of NPAS helicopters passed to Airbus under a national contract. The Met obtained maintenance approvals in 1980 when it first began operating the Bell 222 and retained them over the next 35 years as it moved on to the AS355N and then ultimately the EC145 fleets. The in-house engineering team was managed by a maintenance manager and included a number of EASA B1 licensed engineers who worked alongside an EASA B2 licensed avionics engineer, unlicensed engineers and support staff to deliver up to 3,300 flying hours across the Met's fleet annually. What this meant was that one of the Met's three helicopters was almost always in deep maintenance while the

other two were either on line delivering the air support service or available as a spare airframe. Coordination of the maintenance activity and operational flying was crucial to ensure that this small and exceptionally efficient maintenance team was not overrun with multiple helicopters all requiring complex base maintenance at the same time. It is important to ensure that the daily rate of flying is managed, and although it is not possible to truly control this, as demand for air support service can vary day by day, it is possible to plot trends and adjust maintenance plans to ensure that the engineering team is ready to receive the helicopter when maintenance is due.

The daily rate of flying required is at the core of any maintenance plan. The maintenance inspections and part replacements are governed by the number of hours flown or calendar time elapsed. The key issue is that all maintenance activities must be captured and completed in the most efficient way. The main factor limiting the availability of a helicopter to fly is the maintenance required at each interval and the ability of the MRO to undertake that maintenance. If the operator requires a helicopter to fly an average of five hours every day this means that every ten days the 50 flying hour maintenance activities will be required. Then after another ten days the 50 flying hour and the additional 100 flying hour maintenance activities will be required. The operator needs to make the helicopter available for this maintenance for a sufficient period of time every 10 or 20 days. When the helicopter is in maintenance it is off line and not available for flight, and the operator needs to plan for this unavailability. This is relatively simple for the smaller maintenance operations, but larger base maintenance activities, such as 400 or 800 flying-hour inspections, can take many weeks to complete depending upon the engineering personnel resources available. This can cause problems for an emergency service operator, as policing is a 24/7 business and these off-line periods can be inconvenient; they are, however, unavoidable.

Scheduled maintenance activity is broken down into a series of checks, inspections and part replacements, each of which must be completed in order and signed off by

an approved person. It is possible for an experienced MRO to determine how many man hours each inspection should take. For example, a 400 flying-hour inspection may typically take 200 man-hours of maintenance activity to complete. An MRO with two engineers working a 40-hour week would have around 72 man-hours of maintenance capacity available each week after taking breaks into account. A 400 flying-hour inspection would take that MRO around three weeks to complete assuming that everything went to plan and no unscheduled or unexpected issues were encountered. It is possible to shorten the maintenance period by having a greater number of engineers available, working two shifts a day or at weekends. However, this is only possible up to the point where the activity is limited by access to the airframe. There is a limit to how many people can physically and safely work on a single helicopter, and it is also often the case that activity is required, such as removing a certain part, before a specific inspection is possible. This is known as the maintenance opportunity, and the aim is to optimise the maintenance plan to take into account manpower resources and maintenance opportunities available. It has often been said that planning and scheduling maintenance activity and resourcing is a fine art, and invariably in the police world it is the operator who scuppers all the best laid plans.

Beneath the panels of the EC145

The EC145 – now known as the H145 – is effectively a BK117-C2 variant of the original BK117-C1 helicopter. The avionics and other components have been updated but much of the mechanical side of the helicopter is unchanged. It is worth noting that when the helicopter was converted into a police role equipped helicopter around 100kg of additional wiring was introduced along with other modifications, all of which need maintenance. It requires maintenance at 30-, 50-, 100-, 400- and 800-hour intervals, with the 400-hour maintenance referred to as an intermediate inspection and the 800-hour referred to as a periodical inspection, which must be completed

LEFT The main rotor head close up, with rotor blades attached.

every 800 flying hours or 12 months, whichever is sooner.

The rotor head is made from solid titanium and is attached to the rotor mast with 12 titanium studs and special retaining nuts. The strength and security of this assembly is reassuring and the four rotor blades are attached to that. This rotor head assembly has a part life of 25,000 flying hours and the main rotor shaft 30,000 flying hours, meaning that in many cases these components will long outlive the ownership and operation of any given helicopter. In policing terms the rotor head assembly would need replacing after around 25 years, which is typically longer than the police would own and operate a helicopter. The reality is that the manufacturers consider this to be a strong and durable part that should almost last the life of the airframe.

There were two engineering hangars at Lippitts Hill. The north hangar had a small crane that could be used to lift out engines or the rotor head and was generally used for deep maintenance activities. Helicopters would

ABOVE The main rotor head and engine bay during a ground run following base maintenance. The engineers complete a number of ground runs to check for leaks before moving on to the complete track and balance activities. To complete the ground-based power assurance checks, several hundred kilos are added to the helicopter. These custom-built weights are secured into the floor runners of the helicopter cabin.

RIGHT One of the Arriel 1E2 engines from the EC145 during installation.

RIGHT G-MPSA, one of the EC145 fleet, undergoing maintenance in the north hangar surrounded by custom-built engineering platforms that facilitate safe and easy access for the engineers.

RIGHT Completion of periodical inspections requires the helicopter to be stripped down to a significant extent. This base maintenance activity can take several weeks to complete. Here one of the EC145 fleet is in base maintenance at Lippitts Hill.

LEFT One of the EC145 fleet with all of its internal panels removed during base maintenance at Lippitts Hill.

RIGHT The titanium main rotor head removed from one of the Met's EC145 helicopters for inspection.

BELOW One of the Met's EC145 helicopters undergoing base maintenance at Lippitts Hill. Note the bright red cockpit voice and flight data recorder (CFDR) often called the 'black box'.

BELOW RIGHT One of the Met's EC145 helicopters undergoing base maintenance at Lippitts Hill. Note that the rotor blades and head have been removed and both engines are visible, along with the cockpit voice and flight data recorder.

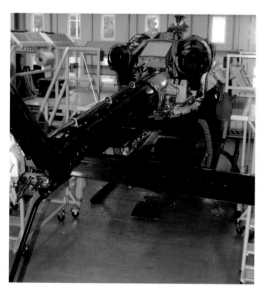

disappear into the north hangar and literally be taken to bits over a period of weeks. The south hangar was much bigger and had a larger crane capable of lifting the entire helicopter airframe. This was used a number of times when the removal of the undercarriage for inspection was mandated in the maintenance schedule. It was on inspections like these that wear and tear was found, and this frequently resulted in a new repair scheme from the manufacturer as they had not yet come across that specific issue. There can be advantages in being a type leader as far as flying hours are concerned: the manufacturer may choose your airframe to trial modified procedures and these can be of benefit. However, there can of course be a cost, as new and previously unforeseen issues crop up from time to time and the manufacturer has to go back to the drawing board for a fix.

The EC145 is powered by two Turbomeca Arriel 1E2 engines. These are impressive looking things when you see them on an engineering bench as they are prepared for installation. The engines are manufactured by Turbomeca (now Safran) and covered by a PBH contract. Turbomeca assures the supply of a serviceable engine at all times, and when one of the engine

modules reaches its timed life a new engine is supplied ready to be fitted. Some MROs have approval for engine module changes, but this was never something that was possible at Lippitts Hill, largely down to the lack of suitable workshop space. As the engine comprises a number of separate modules they can all be lifed at a different rate; the Met contract was always that the minimum life remaining on a module had to be 500 flying hours, but this was almost always exceeded. The main challenge was the engineering time needed to complete an engine change, set the engine up correctly, complete power checks and then schedule test flights. This all took valuable time, and avoidable engine changes were not welcome.

The EC145 has a VARTOMS Type 1. This system is extremely complicated to set up, and over the years the Met engineers became very efficient in setting a multitude of different parameters to get it working correctly. In simple terms the VARTOMS aimed to match the torque delivered by both engines and vary the rotor speed based upon a number of variables, such as airspeed, altitude and outside air temperature. It did this by setting the torque of the right engine with reference to the left engine, to provide maximum efficiency at all times. A CAT A mode can be selected, which raises the rotor speed N^R slightly to provide extra power to maintain CAT A performance. The EC135 has a similar feature which is sometimes referred to as a High N^R setting. The CAT A switch is located on the instrument panel below the CAD and is selected for take-off and landing phases or as required by the pilot. After significant maintenance, such as engine or gearbox changes, the helicopter is put through an extensive ground running programme before moving on to flight checks, which involve climbing to altitude, retarding one engine for single-engine power assurance and so on. The Met's helicopters were designed with this ground running and rotor balancing in mind, and the control cables were integrated into the airframe in order to make it quicker and easier to plug in track and balance equipment.

In addition to the complex avionics suite present on the EC145, the addition of police mission equipment added significant extra complexity and challenges for the engineering

BELOW The No 1 port side engine cowling is removed on the EC145, exposing the Turbomeca 1E2 engine.

and avionics staff who were charged with keeping the helicopter flying and fully operational. There are a number of avionics bays within the EC145. One is under the belly panel to the front of the helicopter, and this is where the Skyquest VMS is installed in order to feed all the police mission displays. The rear avionics rack is located above the clamshell doors and is able to fold down supported by two hydraulic rams for maintenance activity. This rear shelf was modified extensively to house the four police Airwave radios, NAT intercom amplifiers, the TCAS processor and the Garmin radios and navigational aids. The additional weight of the rear shelf equipment is such that great care has to be taken to ensure that it is supported during extensive maintenance activities.

This whirlwind tour of the EC145 is deliberately simple. Any engineering staff reading this book would no doubt be able to talk for hours regarding the trials and tribulations of maintaining such a sophisticated helicopter. The engineering staff work tirelessly to ensure that they deliver a serviceable helicopter for the pilots and TFOs to fly operationally. The achievement of any ASU is measured by the sum of its parts, and there is no doubt that the engineering team forms the very foundation of any unit. Without maintenance there is nothing to fly and with nothing to fly it is not possible to deliver any form of service. For 365 days a year and in all weathers engineers work to

ABOVE LEFT The centre console. The CAT A switch has been pressed to illuminate CAT A in green text. This switch is located just below the caution advisory display and is used to increase rotor speed, thereby maintaining CAT A performance.

ABOVE The rear avionics shelf is opened and supported for ease of access on two hydraulic rams, similar to those on a vehicle tailgate. This has been extensively modified on the EC145 to accommodate additional avionics equipment.

BELOW A close-up look at the rear avionics bay. The four police airwave radios are located on the left-hand side, with the TCAS processor in the centre and Garmin ADF and DME radios towards the right.

ABOVE LEFT The tail rotor cowling showing its clear warning labels; the tail rotor under power is almost invisible and highly dangerous.

ABOVE In an effort to enhance safety, yellow tips make it easier to pick out the tail rotor. Walking into rotors under power is one of the major hazards that all crews are reminded of regularly.

LEFT The tail rotor assembly with the tail rotor cowling removed. The tail rotor control actuator and blue pitch links can be clearly seen. The tail rotor is one of the most crucial components on a helicopter, as without it the maintenance of stable flight is very challenging if not impossible.

RIGHT The EC145 at Lippitts Hill during post-maintenance ground running.

keep the helicopters flying, and ensure that the sophisticated suite of avionics and mission equipment is operating as it should.

Dispatch and flight following

The management of air support demand has always been a major challenge. In the years before NPAS, when forces typically had their own helicopter or they shared one with one or two other forces, it was possible for the TFOs to monitor two or three force radios for tasking and to be tasked directly by force control rooms. The Met was the only force in the UK to have its own dedicated ASU control room, and the requirement for this was driven by the sheer volume of demand received from London's 32 individual borough control rooms. The creation of NPAS changed this drastically, as there was now a national service that received tasking from 43 forces plus British Transport Police, creating a need to coordinate and manage the calls for service. NPAS receives around 1,000 calls for service every week from the forces in England and Wales. Its capacity, just as it always has been for air support, is limited by a mix of flying hours, resources and money. The average police helicopter will fly around 900–1,000 hours per annum, and maintenance contracts and resources are established to deliver this rate of flying per airframe per year. In order to deliver a greater number of flying hours it is necessary to either use more helicopters or increase the maintenance provision; both add significantly to costs. Flying capacity is therefore finite and needs to be managed to ensure that air support capability is available as and when it is most needed. Calls for air support service are carefully assessed to establish that they are viable, legitimate and meet the assessment of threat, harm, risk and vulnerability – thereby warranting a deployment. The Met's ASU established a control room facility at Lippitts Hill in the late 1980s, and in those days the requirement was that a sergeant had to sit in the control room whenever the Bell 222 helicopter was flying. As the use of air support increased the demand turned out to be so great that a need to filter and manage demand became crucial, and the ASU would eventually become

ABOVE Maintenance activity is not a fair weather business, and engineers routinely brave the elements to complete ground runs and checks and deliver the helicopter back into operational service. Here one of the Met's EC145 helicopters is undergoing post-maintenance inspections after a ground run in the snow.

CAD live. This meant that they had real-time access to the Met's CAD system that is used to manage all policing deployments in London.

The task of managing air support demand falls to the NPAS operations centre, which is based in West Yorkshire. Its control room receives calls for service from forces and assesses their viability before dispatching the closest or most appropriate asset. The crew is then notified as soon as possible of the type and location of the call along with the radio talk-

LEFT The Met's dedicated air support control room at Lippitts Hill in the 1980s. The requirement in those days was for a sergeant to be present in the control room whenever one of the Bell 222 helicopters was flying. Police Sergeant Tony Mepham is pictured here.

group; this allows a rapid take-off and arrival on scene at the earliest possible opportunity for those calls that require an urgent response. Prior to NPAS the various helicopter units were dispatched by their force control rooms, within which a force incident manager would authorise and request the assistance of air support. Typically ASU bases would be policing only two or three force areas, and as such they were able to monitor the operational radios for viable tasks, often being one step ahead of the control room. The move to a national borderless service with the creation of NPAS meant that it was no longer considered viable to allow individual forces to request and dispatch the ASU service directly; some form of national oversight and management was needed.

The Metropolitan Police with its 32 London boroughs had always struggled with the volume of demand for air support. In the mid-1990s the demand was growing and the Met's command and control hierarchy were persuaded to establish a dedicated air support control room based at Lippitts Hill. The ability to connect the 32 local control rooms directly with air support via the CAD system was pivotal in the success of this. Over the next 20 years or so the control room functions increased, as did the technology. The room was staffed 24 hours a day every day by a single member of police staff, known as the air operations officer, who

undertook the command and control function managing all helicopter deployments and became considered to be the fourth member of the crew. The primary role of the air operations officer was to assess the viability of any given request for service and to task the aircraft as appropriate. However, this officer was much more than a dispatcher – and the control room became the glue that held together the whole unit. The Met's ASU control room handled around 60 requests for air support every day and filtered the requests down to dispatch to an average of around 20 tasks per day.

The control room staff at Lippitts Hill became highly skilled in the management of air support demand. It was packed with technology, including weather monitoring equipment, video walls, downlink receiver control, tasking computers, Met CAD and a multitude of police and other radios. There are so many variables in relation to air support that careful and considered management is required, and this is achieved through a single point that has an overall picture of what is going on, what is coming up and the capacity of the unit to respond.

Managing capacity – Air operations officers play a crucial role in the need to manage the rate of flying, as the cost of this is high and the capacity to fly is limited. If, for example, a helicopter needs maintenance every 400 flying hours and that 400-hour maintenance operation takes two weeks to complete, then the daily rate of flying is important. If the unit were to fly four hours per day, then every 100 days the helicopter would be off line for two weeks. If the rate of flying is eight hours per day then the maintenance comes around in half the time, every 50 days (seven weeks). It is therefore crucial that the MRO is consulted daily regarding maintenance requirements and the daily rate of flying is monitored closely. This ensures that the helicopter does not run out of flying hours capacity before the maintenance organisation can complete the maintenance. If this happens the helicopter is grounded, and there will be no air support available.

Assessing viability – With limited flying capacity available, it is essential that a crew is dispatched to tasks that are viable and where they are likely to reduce risk and add value. This assessment of task viability was primarily undertaken by the air operations officer, and they did this through careful liaison with the requesting control room and consideration of the individual circumstances present. A request for service in relation to a suspect search has so many variables: if the request is recent, the suspect has been chased by police, a description obtained and their potential location identified, the request may be viable. If, however, the suspect hasn't been seen plainly, their location is not clearly identified or the time elapsed is great then the chances of air support success are reduced and the task is unlikely to be viable. If you then add in risk factors such as the presence or involvement of weapons or violence, intelligence to suggest their involvement or the proximity to height, railways or water, the request will increase in viability. It is certainly not an exact science and there is no substitute for training and experience, with air operations officers quickly becoming very efficient and effective.

Flight following – The safe operation of an ASU is crucial. NPAS is required to maintain a status awareness of all their helicopters at all times, and this process is known as flight following. In the Met's control room flight following one or two helicopters was achieved

by the air operations officer, who was able to communicate with the helicopter pilot and crew via police or company radios. In addition the air operations officer would routinely monitor the working talk-group of the borough where the helicopter was working and could hear the crew communications and the progress of the task. It was also policy to downlink live video imagery throughout the task, and this was displayed on the video wall in the ASU control room. The air operations officer routinely supported the crew, ensuring that they had all the information needed to complete the task and on occasion assisting them with the assessment of the live video. On more than one occasion the control room spotted a suspect on video downlink who had not been picked up by the crew. It is because of this interface between the requesting borough and the helicopter that the air operations officer was frequently referred to as the fourth crew member.

Police air support can fly in very poor weather conditions, but unlike some SAR assets they cannot fly if the cloud base is too low or if the visibility is too poor. The weather limitations are exceptionally complex and are the subject of a whole lesson on the TFO training course. There are differing limitations by day and night and at the base, en route and on task, all of which must be considered. The police helicopter pilots will not fly from A to B in a

ABOVE The control room at Lippitts Hill overlooked the apron and had unhindered sight of the helicopter landing pad and clear area. A single air operations officer was responsible for assessing the viability of requests for service and for tasking the helicopter over 20 times on an average day.

straight line, and often when they take off they have no idea exactly where they will end up other than that they intend to return to base at the end of their flight endurance; this is driven by fuel uplift. The air support base has weather monitoring equipment to measure temperature, air pressure, visibility, wind direction, wind speed and cloud base. This information is collected by a weather station near the clear area and helipad and is displayed on monitors in the air support control room. From Lippitts Hill as a very basic summary, the visibility by day must be 1km or greater, and by night this increases to 3km or greater. The helicopter must operate at least 50ft below the cloud and at least 300ft above the ground by day and 600ft by night. There are further complications when it comes to proximity to buildings. It is therefore no exaggeration to say that staying legal is complicated, and changes in the weather can have a dramatic impact upon air support operations. It is worth noting that 1km of visibility for a helicopter travelling at 120kts plus and flying at 300ft above ground level in and among the buildings and cranes that stand at heights of up to 1,000ft in London is no joke, and police air crews take this very seriously to stay safe at all times. The role of the control room is crucial to assist with this, as it can monitor the weather in real time, warning of deteriorating visibility or reducing cloud height. The worst case scenario for the helicopter is that it is unable to return to its base and has to either divert to an airfield or simply find somewhere to land and wait for the weather to clear. The aim is to fly from and return to the base, and chasing gaps in the weather is never a good idea.

The Met's ASU control room was retained alongside the NPAS operations centre for around three years after the Met joined NPAS. Eventually the costs of retaining a satellite control room led to its closure, and the whole of the NPAS tasking and flight following is now undertaken by the NPAS ops centre in West Yorkshire. This is resourced so that each region of the country has a dedicated dispatcher who is responsible for assessing the viability of calls for service within that region, tasking the closest or most appropriate NPAS asset and flight following. The link to the maintenance management team means that the ops centre retains an oversight of the flying hours available and the status of each base and helicopter. Flight following is achieved via GPS tracking of police radios carried in each helicopter, and helicopters' positions are displayed on large electronic maps around the control room.

Media and publicity

Police ASUs have always attracted a great deal of attention both from the public and the media. It is hoped that this book will satisfy some of the fascination that people have regarding what that helicopter is doing when it orbits over their house at night. In an

BELOW A local newspaper article from November 1980 about the formation of the Met's ASU.

GAURDIAN & GAZETTE NEWSPAPERS
(WALTHAM FOREST)

28ᵗʰ NOVEMBER 1980

CHOPPER SQUAD TAKES TO THE AIR

THE Metropolitan ice has become the last British police force to buy its own helicopter. At Lippits Hill, High Beech, on Wednesday.

The new Bell 222 helicopter was officially "unveiled" by Home Secretary Mr William Whitelaw who launched the newly-constructed base for the Metropolitan Police Air Support Unit.

The American helicopter costs £600,000 and the sophisticated police equipment adds a further £150,000.

Some of this includes Marconi Avionics Heli-Tele camera system, the Decca Tans F12 computer display navigation system, integrated communications system, a hoist and has space to take two stretcher cases.

As a safety precaution the Civil Aviation Authority insists on twin-engined machines capable of flying on a single engine.

"Buying the helicopter will save us a lot of money in the long run. We maybe in profit within three years," said Assistant Commissioner John Wilson.

The success rate of police use of helicopters on burglary and robbery investigation is high. The advantage is that the crew can search areas inaccessible to police ground units, direct and re-direct officers, police vehicles and equipment rapidly.

Traffic and public order situations, like demonstrations, can be easily assessed from a helicopter. Other areas of helicopter use would include crime investigation, protection, escort of high-value loads and co-ordination of police action at scenes of fire, flooding or other major disasters.

The helicopter, with its distinctive orange and white markings, will retain the familiar police radio call sign of India 99.

The 222 has a cruise speed of 155 mph and is permitted to fly as low as 500ft. There is currently no equivalent British helicopter.

Trained police dogs are also being taken into the air. The police hope to start on their training shortly.

The Home office is currently experimenting in picking up number plates from the air, at Sandridge. It is claimed that they have a 60 per cent success rate so far.

● The new Bell 222 helicopter

● Police chiefs and crew members with the new helicopter (1087).

effort to lift the lid on air support in the UK, the ASUs in South Yorkshire and the Met allowed the TV cameras to film with them, fly-on-the-wall style for a period of several months. The two *Skycops* series were aired on BBC One at prime time, and took the viewer inside air support to meet the crews and take viewers along for the ride on several missions.

Another first for police aviation was a three-hour radio broadcast that brought Radio 5 Live presenter Steve Nolan to Lippitts Hill so that he could broadcast his Friday evening show from the base. The show was a mix of pre-recorded pieces, live studio discussion, interviews and a telephone phone in. The highlight of the show was the first ever live radio link up with an operational police helicopter as it worked. Following on from this, a further foray into radio was undertaken as a pre-recorded Radio 4 series *Night Visions*. This was a poetic and beautifully descriptive journey into the work of air support at night over London. Poet Paul Farley was the presenter, and he crafted a piece called 'The Asset' to accompany the series.

Other TV and radio appearances also followed with *Police Camera Action*, *Crimewatch Live*, *London Tonight* and LBC radio to name just a few. The crews working in air support were generally not shy, and they enjoyed their five minutes of fame as these TV and radio appearances were aired. There is absolutely no doubt that the noise of a police

LEFT A newspaper article from the London *Evening Standard* about the use of Twitter to tackle noise complaints.

BELOW A promotional flyer for the *Sky Cops* BBC One TV series.

helicopter working at night is frustrating and inconvenient, as it wakes whole households. In order to address this, the Met took to Twitter with an account that has now morphed into @NPASLondon. The basic idea was to engage with the public by explaining what the helicopter was doing, as far as was possible, as soon as it finished any given task. The public were encouraged to follow Twitter so they could quickly and easily learn what the helicopter was doing. The move was a great success, and the sharing of exclusive imagery, behind-the-scenes information and cockpit pictures proved really popular, as did the ability to ask the crew a question and receive a personal answer. The number of noise complaints was cut, and soon the account had over 160,000 followers. Today NPAS has expanded this, and each base has its own Twitter feed today.

LEFT The author and Sergeant Andy Hutchinson answer questions from the public during a phone-in as part of a live Radio 5 Live broadcast.

RIGHT PC Tony Donnelly braves live television as he records a short interview for the BBC *Crimewatch Live* series; this episode was filmed at Lippitts Hill.

Index

ABOVE An anniversary version of the Met's air support crest was produced in 2010 to celebrate 30 years of air support. It featured all three helicopters that had been in the Met fleet: the Bell 222, AS355 and EC145. It was not used, however, as the colour scheme on the EC145 was not correct. *(Hugh Dalton)*

BELOW The first of the redesigned ASU logos featuring the AS355 helicopter.